Living Coral Reefs of the World

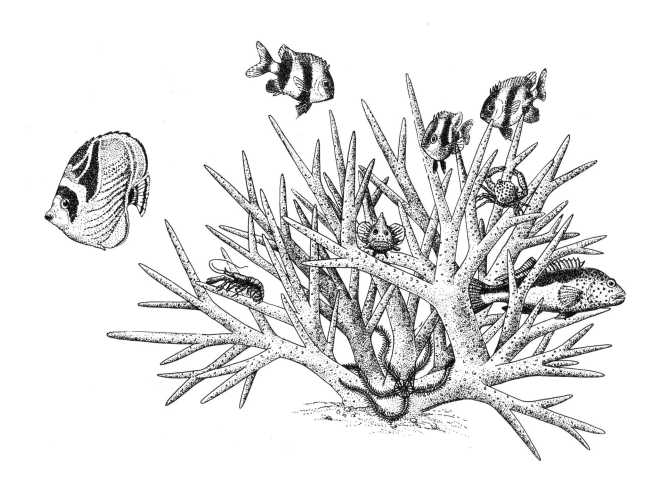

Dietrich
H. H. Kühlmann

Living Coral Reefs
of the World

ARCO PUBLISHING, INC.
New York

Translated from the German by Sylvia Furness

Published 1985 by Arco Publishing, Inc.

215 Park Avenue South, New York, NY 10003

© 1984 by Edition Leipzig
Design: Eveline Cange
Drawings: Britta Matthies

Library of Congress Cataloging in Publication Data

Kühlmann, Dietrich.
    Living coral reefs of the world.

    Translation of: Das lebende Riff.
    Bibliography: p.
    Includes index.
    1. Coral reef biology.   I. Title.
QH95.8.K8413 1985 574.909'42 84–14515
ISBN 0-668-06327-0

Printed in the German Democratic Republic

Contents

5

Foreword

A book has been written about a curious, secret, exotic world within the ocean, about coral reefs. The account is in no way romanticized, yet full of life and beauty. But I wanted it to be more than the description of external phenomena, for it is my aim to foster an understanding of the potentialities and conditions of existence of this biotope that has been created by the activities of the organisms themselves. To this end, it has been necessary to examine the relationships of cause and effect. The material, then, demands a degree of concentration and endurance on the part of the reader. Let him not be disheartened, for he will be enriched by it. He will come to see the coral reef and its creatures with new eyes; he will not merely gaze at it in wonder, saying "What a wealth of amazing creatures nature has produced!", but rather will question *how* this comes about. And he will seek for ways in which mankind can ensure the preservation of the coral reef biotope, in face of the dangers that threaten it. This is my hope.

The book is the combined work of all those who, asking "How?" and "Why?", have tirelessly applied their intellect in an attempt to detect the inmost force, "Which binds the world, and guides its course" (Goethe)—in this case, the world of the coral reef. Rightfully, hundreds of names should appear on the title page—many more than could be included in the bibliography, which, for reasons of space, had necessarily to be limited. Many people contributed directly to the production of this book. First of all I would like to mention my wife and colleague, Karin Kühlmann, who prepared specimens collected during my travels, wrote up manuscripts, prepared lists and read proofs. The material help and organizational support of colleagues in the Ministry of Education in the German Democratic Republic and the Humboldt University in Berlin, in particular K. Baltruschat, B. Hölzer, E. Mehls, W. Riedel, Chr. Schmidt-Renner, L. Ulinski and U. Zimmermann, and my fellow members of staff in the Museum of Natural History in Berlin, K. Senglaub, K. Deckert, R. Daber, W. Vent, G. Hoppe and H.-J. Hannemann made it possible for me to carry out underwater research in the tropics. Valuable assistance came from many parts of the world: from D. Guitart, P. Duarte-Bello and O. Santanas in Cuba; N. K. Panikkar †, N. B. Nair and S. Mahadevan in India; J. Thomson and his wife, R. Endean and K. Rohde in Australia; A. Antonius and his wife, R. Ginsburg and W. Adey and his wife in the U.S.A.; R. Dill and H. Tonnemaker on the Virgin Islands; J.-P. Chevalier †, B. Salvat and H. Zibrowius and his wife in France and finally G. Vergonzanne, J. de Vaugelas, G. Richard and A. Hauti on Moorea and in the Tuamotus. Travelling the Atlantic, the Red Sea and the Indian Ocean in ships of the Rostock Shipping Line, I was supported with help and advice by H. Schröder, G. Engel, M. Dux, A. Herold, J. Voss, K.-H. Zink, H. Bohk, J. Mokelke, S. Rathmanner, D. Strulik and my assistant and friend, J.-S. Lüdecke. I wish to thank A. Kupke for the underwater electronic cameras. Photographs used in the book were put at my disposal by R. Bak, H. Fricke, J.-G. Harmelin, K.-G. Heckel, S. Johnson, P. Playford, K. Rabe, E. Reese, H. Schöne, H. Schuhmacher, D. Strulik and K.-H. Tschiesche. My father, Hermann Kühlmann †, J.-P. Chevalier, F. Fosberg, H. Fricke, O. Kinne, M. Pichon, G. Pillai, J. Randall, G. Scheer, H. Schuhmacher, D. Stoddart, J. Veron, C. Wallace, J. Wells, M. Mijsman-Best and others gave valuable assistance in assembling the necessary technical literature.

Further thanks are due to Frau B. Matthies, for her skill and patience in making the sketches, Frau E. Cange, who undertook the graphic design and Mrs. Furness for her excellent translation of the German text into English. I am grateful to the publishing house of Edition Leipzig for their generous support.

Inevitably, many of those who helped me have not been mentioned. I thank them all for the interest they have shown and their unfailing help. I can only hope that my work may prove some small justification for so much friendship, understanding and cooperation.

Dietrich H. H. Kühlmann

Introduction

to a fantastic underwater world

Our earth is a huge sphere of rock like thousands of other stars. But it is probably more beautiful and more richly varied than they are, because it possesses an abundance of water in the form of the oceans, that provide particularly favourable conditions for living organisms to develop and evolve. Frequently it is plants that determine the character of the earth's landscape—savanna tracts, steppe lands and forests with their green vegetation, with fruits, butterflies, birds... Similar "landscapes" exist as jungles of water milfoil and pondweed in inland waters and as seagrass meadows and seaweed forests in the oceans. There is, however, a unique biotope of unparalleled beauty that has been created by animals: the coral reef. Lime-depositing organisms shape these underwater mountain ranges that house an almost indescribable wealth of living creatures.

Yet the organisms that have built the reefs, the coral polyps, also exhibit "plant-like" characteristics. They are firmly attached to the substratum and form rigid shrub-like or fungiform growths. When the coral polyps extend their circlets of tentacles into the open water, the "bushes" have the appearance of being covered with flowers. Today, a considerable amount is known about these curious creatures, but by no means have we been able to wrest from them all the secrets of their fascinating life. Ultimately they produce calcium in such quantities that they form reefs. Stony corals, then, do not simply cover the sea

bed as grasses carpet the soil of the steppes and trees the floor of the forest, but by aggregation they elevate the very surface on which they have settled. Beneath this purely biological concept lie harmonized forces of such vast potential that the biosphere of the coral reef has already been able to exist for more than 400 million years.

In earlier times, when scientific researches were not dictated by prosaic economic considerations to the extent that they are today, men of science exploring tropical oceans often expressed enthusiasm for the wealth of form and colour in the organisms they found living there. In the diary of his travels written about a century ago, Ernst Haeckel observes: "It is not really possible to describe a coral reef; to be fully appreciated, it has to be seen."

In those days, a visit to a coral reef was an arduous and dangerous undertaking. Explorers often ran the risk of tropical disease. Between 1820 and 1826, when Ehrenberg, the founder of modern micropalaeontology was travelling in the Red Sea region, nine out of the eleven members of his expedition met their death. But there have always been men who, in response to some irresistable urge have returned again and again to unexplored regions of the world, to icy polar zones, dense jungles, parched deserts. No dangers could deter them. But the underwater world remained closed to explorers for a long time. To scientists working from ships, the ocean

yielded up its secrets only in fragments. Only the development of effective diving apparatus in the last few decades has made it possible to penetrate even this, the most precious and beautiful of ocean biotopes, the coral reef.

The coral reef regions of the world cover an area greater than that of Asia, Europe and Africa together. Many of the reefs rise from the sea as islands and are inhabited by man. Lying off the coast, they protect shore dwellers from the ravages of the sea, while the creatures living in the reefs often provide an important basic food. The few decades during which they have been extensively investigated have led to discoveries of considerable economic benefit: mineral oil, building materials, medicines... Utilization of the coral reefs has already started—and started much too soon, since neither social acceptance of responsibility nor the state of scientific knowledge is sufficiently comprehensive to safeguard the reefs from harm. The serious increase in environmental pollution over the last few decades has not spared the coral reefs. From this point of view, examination of the reefs has become an urgent necessity. Today, Haeckel's words quoted above might well be modified to read: To appreciate a coral reef, it is not sufficient to see it, but the cause and effect of its development and existence must be understood in order to ensure both its proper use and its preservation. It is to this purpose that I dedicate this book.

Corals

in history, corals today

Anyone seeking a better understanding of the present would do well to cast a glance at the past. We too shall be nearer solving the enigma of coral if we know how man first discovered it, how he regarded it in classical antiquity and in medieval times, and how finally, at a quite late date, he recognized its true animal nature. But the path that led to this was a long one, with its beginnings in the shadowy world of thousands of years ago.

Whereas the men of the northern shores remained long in ignorance of corals and of the abundantly rich world of the coral reef, the peoples of the tropical seaboards had long been familiar with them, because thousands of years before the Christian era, Oriental and African peoples settled on the shores of the Red Sea and of East Africa, while Asiatic and Australian tribes travelled across vast distances on flimsy rafts and took possession of the islands of the Pacific. The coral world was part of their life. How was it possible that in their frail craft and without any accurate knowledge, they were able to master the reefs and breakers, and wrest from them their food?

Knowledge of corals may have come first to the Orient 6,000 years ago and have been carried from there to Europe; even at that early date, women of rank in Babylon already wore pearl jewelry. Slaves would dive to gather from the coral reefs of the Arabian Gulf the large, dark shells that were sometimes found to

contain a pearl. No doubt they recounted their experiences below the surface of the ocean, and it may be that their words were the starting point for the circulation of stories about a world of corals, which became increasingly inaccurate and nebulous with increasing distance.

Some time between 1000 and 500 B.C. the sage, Zoroaster, lived in West Irian. The first reports of the use of precious red coral as adornment and as ritual objects are attributed to him. Theophrastus, the Greek philosopher who studied under Aristotle and laid the foundations of the science of botany, speaks of red coral in his works written in the 4th and 3rd centuries B.C. In the early years of our era, Ovid celebrated in verse the hero Perseus who cut off Medusa's head and freed Andromeda:

...he washes his victorious hands in
water drawn for him;
and that the Gorgon's snaky head may
not be bruised on the hard sand,
he softens the ground with leaves,
strews seaweed over these,
and lays on this the head of Medusa,
daughter of Phorcys.
The fresh weed twigs, but now alive
and porous to the core,
absorb the power of the monster and
harden at its touch and
take a strange stiffness in their stems
and leaves.
And the sea-nymphs
test the wonder on more twigs
and are delighted to find the same

thing happening to them all;
and, by scattering these twigs as seeds,
propagate the wondrous thing
throughout their waters.
And even till this day the same nature
has remained
in corals so that they harden when
exposed to air,
and what was a pliant twig beneath
the sea is turned to stone above.

It is not clear of which of the corals Ovid is speaking; since their true nature was still unknown, it was assumed at that time that the calcareous skeletons did not harden until they came in contact with air. But his reference may have been to the horny corals in the Mediterranean and which are elastic under water, but out of it, become dry and rigid.

In classical antiquity, supernatural powers were ascribed to coral:
"It protects boat and roof
from lightning, whirlwind and
tempest...
It also wards off demonic shades
and the terrors of Thessalian
witchcraft..."

Whereas the Gauls ornamented their helmets, shields and swords with red coral, the Romans attached little value to it, to judge from Pliny the Elder, who lived in the first century A.D. He makes only fleeting mention of it in his comprehensive *Naturalis historia*: "Our corals are as highly prized in India as Indian pearls are here; it is simply the opinion

9

of the people that determines the value of things." Firm trading links existed at that time between the Romans and the peoples of Asia. At the beginning of the fifteenth century, a Chinese author, depicting Calicut as the trading centre of India, wrote of it, "Here there is pepper, attar of roses, pearls, incense, ambergris, coral...". Oriental priests wore coral ornamentation as a protection against danger, Chinese mandarins as a token of rank.

Pliny recommends the medicinal use of red coral in the treatment of many illnesses and, taken pulverized in wine, as a sleeping draught. The wine may well have proved an effective soporific, with or without the addition of powdered coral. And there is surely nobody today who still believes that little hands fashioned out of coral are a protection against the *iettatura,* as the Italians call the spell of the evil eye, for belief in "evil eyes" and "witchcraft" is declining steadily in face of man's growing understanding of a rationally explicable natural world. Whether the wearer is spared headaches or not, the red coral is in no way responsible. In some regions, it is held to be a protection against impotence. But there is one instance in which its therapeutic effect cannot be denied, namely when it is given in powdered form together with Vitamin D to teething infants. Since it consists of calcium, it can help the growth of teeth and bones. But it is generally known now that for many years, the pharmaceutical industry has been able to supply much cheaper and more effective products for this purpose. So its miraculous power is gone, and it is the beauty of the red coral alone that gives it its due value today.

Red corals

The Mediterranean once concealed a rich treasure within its depths—precious red or rosy corals. Today, they have become a rarity after thousands of years of wasteful exploitation. But the diver fortunate enough to come upon them cannot fail to be enchanted by their beauty. From an overhanging rock face, plant-like growths branching in bizarre shapes extend downwards into the darkness of the watery depths. Illumined by the diver's torch, they gleam a rosy red. The branches are covered with delicate white "flowers"—animal colonies that resemble plants (Ill. 2).

In 1864, the French zoologist Lacaze-Duthiers observed that a parent polyp releases hundreds of microscopically-small larvae into the water. Their roundish-oval body is densely ciliated, and the regular beating action of the fine hairs allows for locomotion in water. In the course of development, one end of the larva extends and develops into the oral aperture. The body assumes a cylindrical shape, settles and attaches itself to a rock face. Eight feathery tentacles grow round the oral aperture—the structures that lend the polyp the delightful appearance of a white blossom. The function of the tentacles is to catch organic food and direct it into the mouth from where it passes into the digestive cavity that occupies almost the entire hollow interior of the animal. The polyp grows and produces buds that develop into daughter polyps. These are in two different forms: tentacle-bearing feeding polyps and minute tentacle-free respiratory polyps. The latter resemble mere pores, furnished with a large number of vibrating cilia filaments through which a continuous supply of fresh oxygen-rich water is moved. By means of a ramifying system of canals that link the polyps with one another, it passes through the entire colony.

Within its tissue, the small coral colony deposits large numbers of jagged red limestone needles. They serve to stiffen the soft body. Increasing numbers of first and second generations of polyps finally cause the development of a many-branched colony. Individual skeletal needles are no longer adequate as stiffening, even when they are linked together with horny fibres. So over and above the limestone needles in the coenosarc, the coral colony proceeds to amalgamate with many others to form a rigid hollow cylinder. With the further increase in the thickness of the colony, new layers of limestone needles are continually deposited in the form of concentric rings, so that the wall becomes progressively thicker. In this way, a rigid red axial skeleton is formed.

The true home of precious red coral is the Mediterranean. It is also found in certain parts of the eastern Atlantic, such as Cape Verde and the Canary Islands. Usually it grows at a depth of 80 to 200 m, but has also been found at a depth of less than 10 m and more than 300 m. In shallow water, the animal colonizes gloomy grottoes since it prefers darkness. There, the colonies depend from the roof, while at greater depths they stand upright on flat to steeply sloping ground. They always grow in calm water where the brittle, fragile skeletons are not in danger of being destroyed by the waves.

For many years now, coral fishers have been using an apparatus of hardwood rods bound together to form a cross with a diameter of between one and eight metres, weighted by stones or iron sinkers to dredge up corals. It is hung with thick ropes and coarse netting and sometimes equipped with iron prongs and spikes. Working from a boat, the fishermen drag it along the rock floor, intermittently raising and lowering it. On steeply sloping faces, pieces of equipment rather like landing nets with metal teeth and net pouches are used. In both cases, the red corals are detached forcibly from the rock surface and caught up in the net. But many are left behind lying on the sea bed where they perish; since young colonies are also broken away, these methods represent really destructive exploitation. For this reason, the Strait of Messina has been divided into ten zones. Each year, fishing for coral is permitted in only one of them. Although this action provided red corals with an interval in which to re-establish themselves, the extent of destruction since has been such that in many places today, they have been entirely wiped out.

Coral divers work more efficiently. Even by the Middle Ages, naked divers had effectively cleared the coral stocks in the shallow waters and were unable to reach those that were restricted to the greater depths. Today's coral diver is equipped with compressed-air diving equipment. Although a depth of 40 m is considered to be the limit for safe operation, some divers will go to depths of 100 m or more to reach precious red corals. They gather the highly-prized treasures in net baskets which they heave, full and heavy, into the boats. Failure to observe the correct decompression procedures sometimes leads to diving accidents which can cause paralysis and death. But an annual profit of between 50,000 and 80,000 dollars repeatedly induces foolhardy behaviour of this kind. At the moment, the last stocks of coral are still safe, because they are situated at a great depth. But if one day the use of helium-oxygen mixture in breathing equipment were to become established among professional coral divers, the complete eradication of this very beautiful and precious Mediterranean heritage would be only a question of time.

The prices that the coral fishers were paid for their harvest during the many centuries in which they traded, depended upon fashion, and fluctuated with supply and demand. In addition, the quality of the red coral affects the prices considerably. Of the branches up to 4 cm thick, the thickest are particularly sought after for carving. Often their value is reduced by numerous tiny channels that are the work of boring organisms. The colour of the axial skeleton of *Corallium rubrum* (Linné) varies from scarlet, vermilion and matt red to rose pink. In rare cases, it is white or black outside and red inside, or else red striped, spotted or dark red with white patterning. Naturally, less common colours or those which were particularly fashionable commanded higher prices than others. For example, in the second half of the last century, pale pink corals, known in the trade as *peau d'ange* (angel's skin) were particularly in demand. Whereas the latter might cost as much as 400 to 500 francs, other colour variations would fetch 40 to 70 francs, and skeletons with many tiny drill-holes no more than 5 to 20.

But on the whole, coral fishing and processing was a highly profitable trade until the Second World War. After the discovery of the extensive coral reefs at Sciacca in about 1880, the annual landing of red coral in the harbours of Torre del Greco and Porto Empedocle amounted to some 3.5 tons, with a market value of 1 million gold marks, while for the whole of Italy, the total was 5 to 6 tons. At the same time, Spanish coral fishers harvested 1 to 1.3 tons a year. In 1970, one Neapolitan merchant was estimated to have amassed a stock of 1.5 tons of coral.

Once harvested, the red coral was made into works of art and jewelry of high quality and enduring value: necklaces and bracelets, rosary beads, chess pieces and pipe bowls, earrings fashioned of pieces of polished coral set in gold, rings, cameos, brooches and magnificent crucifixes. In decorative works of art, craftsmen combined corals with pearls, diamonds and other precious stones. It was even used to adorn the censer owned by the Dalai Lama in distant Tibet.

The process of working red coral begins with the removal of the live outer layer. After that the skeleton is smoothed with abrasive cloths and oil, then polished on steel. Fine saws cut the branches to the required size. "Sea pearls" for necklaces and bracelets are worked on lathes. With fine cutting chisels, knives and burins, craftsmen carve wonderful figures and ornaments from the precious material. Guilds and schools of coral carving grew up in Naples, Leghorn, Genoa, Marseilles and Paris.

But there came a turning point. At the end of the 19th century, fishing for a closely related species of red coral *Corallium secundum* Dana began off the coast of Japan. By 1907, Japan had produced 10 tons of this almost equally valuable material—sufficient to supply the whole world. Soon the Japanese had developed their own coral industry, and the availability of cheap labour gave them a decisive advantage over their trading rivals in the Mediterranean countries, where coral fishing and coral working declined during the next few decades.

But a second competitor has appeared on the scene in the form of imitation coral. For some time now, an industrial process has been able to produce from other materials imitation coral skeletons so deceptively like real ones that only the specialist can tell whether they are genuine. Might this perhaps preserve true corals from destruction? For who will want to buy jewelry if he cannot be sure that it is real?

Plant or animal?

There can be little doubt that anyone finding a branch of coral washed up on some warm sea shore would, if he were unencumbered by expert knowledge, take it for some kind of plant material. If it were of hard limestone, the idea of a mineral structure might suggest itself. He would certainly not think it could be an animal. The scholars of classical antiquity also classified red corals as plants, even though with a degree of uncertainty, as we see from the name *zoophyta* meaning "animal plant". It is uncertain when this name was first coined. Certainly it was used by the Sceptic philosopher Sextus Empiricus in about 200 A.D. Themistius in the 4th and Philoponos in the 6th century also used it. The term, which at that time was obviously already widespread, remained in use until after the Middle Ages. Other sessile marine organisms such as sponges, moss animals (Bryozoa) and tube worms (Sedentaria) were included in the large group of zoophytes.

In the following centuries, the view that precious corals were plants hardened into an almost incontestable dogma. The tree-like branching, the hard core enclosed in a softer outer layer, the milky fluid that is secreted from a damaged part "exactly in the manner of a fig tree"—all this seemed to justify the classification of these organisms to the plant kingdom. In 1703, Ray described them as "plants without flowers, of a hard, almost stony nature". Finally, in 1701, the Italian Marsigli notified the Abbé Bignon in a letter that he had discovered blossoms on corals. He sketched them and their "blossoms" with great precision. In addition, he noted a putrid smell when they began to decompose, but under the constraint of his time, he dared not pronounce them to be animal. In 1725 he published his observations in *Natural History of the Oceans*.

There were, however, some men of science who thought that since "silver trees" and "lead trees" develop under certain experimental conditions, "coral trees" are likewise purely mineral in nature, particularly since they too come into being in a salt solution, the sea. Thus Boccone in 1670 and Woodward in 1695 classified corals as members of the mineral kingdom. From there, it was a short step: in 1704, Ludius declared them to be "Lithophyta" or "stone plants".

In spite of the excessive dogmatism of those days, a Renaissance that began in the era of a flourishing bourgeoisie brought a new stimulus to the natural sciences. After almost two thousand years of stagnation, the spirit of Aristotle, the search for truth and objectivity celebrated a triumphant rebirth. Every branch of learning expanded and proud self-assurance spoke out from the works of Belon, Rondolet and Salviani. Aristotle himself had assigned an animal-like nature to many organisms growing on the sea bed. In 1558, in the fourth volume of his *Historiae Animalium* and in 1606, in his *Icones animalium in mari et dulcibus aquis degentium,* Conrad Gesner tried to take up a mediatory position more closely approximating to the Aristotelian view. And Rumphius who studied corals during his time on Amboina, recalls in his book *D'Amboinsche rariteitkamer* (A Collection of Amboinian Rarities) certain Indian philosophers who had expressed the opinion that corals were built by small animals.

The fact that for 2,000 years, corals had been classified now as minerals, now as plants, now as animals, indicates not only the lack of knowledge among men of science but also the considerable difficulties that existed in studying this group of organisms and answering an apparently simple question. No proof existed for either one or another of the hypotheses. But this was soon to change.

After inventing the microscope at the end of the 17th century, the energetic Anton van Leeuwenhoek himself discovered freshwater hydra in 1703. He published his findings immediately in Volume 23 of the *Philosophical Transactions* of the Royal Society in London; with the upsurge in the Natural Sciences, scientific journals had come into being.

In 1723, Jean Antoine de Peyssonnel, a French ship's doctor who had taken a considerable interest in the "flowers" that Marsigli had discovered on the branches of red corals, accompanied coral fishers from Marseilles on their dredging expeditions in the Mediterranean. As soon as the dragnet was hauled aboard, he took corals from the nets and placed them carefully in containers of fresh seawater; after several hours, their branches were covered with white "flowers". But at the slightest touch, the latter immediately withdrew into their housing, shortly afterwards to unfold anew. Beneath the outer layer, Peyssonnel observed canals ramifying through the substance of the coral, permitting communication between polyps. After some experiments, he came to the conclusion that corals are animals in nature and used the terms "polypus", "urtica" or "purpurea" to designate the "flowers".

Two years later, he supervised experiments of this kind on the coast of North Africa, observed the movements of the tentacles, and succeeded in preserving specimens of red coral colonies with extended polyps. After this, he included stony corals in his examinations. In the polyps of these corals, which are sometimes much larger, he observed the intake of animal food. These observations convinced him that all coral colonies are produced by the activities of large numbers of zoophyte-animals, a term that would have to be translated as "animal-plant-animals".

In 1726, he sent an account of his observations, experiments and conclusions to the President of the Paris Academy, the physicist Reaumur. Although the latter delivered the paper to the Academy, he found its conclusions so incredible that he discreetly withheld the author's identity in order to save him from ridicule. Of course he refused to allow the manuscript to be printed, wrote to Peyssonnel pointing out his "curious error" and reiterated his own support for the view that corals are plants. Peyssonnel, who was clearly able to appreciate just how stony the path is for a man of science who has reached a conclusion contrary to the established view, was not discouraged, but took up a post as botanist and physician in Guadeloupe, where he continued his study of stony corals.

In 1738, Trembley, an Englishman, confirmed the animal nature of the hydra discovered by Leeuwenhoek. Since coral conforms with it in structure and behaviour, it was also recognized as animal. Even Reaumur who, to give everyone his due, had previously described corals as consisting internally of stone and externally of plant material inhabited by minute parasitic animals, freely admitted his error. Finally, in 1752, a short extract from an extensive manuscript by Peyssonnel containing a wealth of new observations and astute conclusions on the biology of phytozoa or plant-animals was published in Volume 48 of *Philosophical Transactions*. After quarter of a century of neglect, the scholar had at last received the recognition he so richly deserved for having discovered and established the concept of corals as animals.

But old ideas tended to persist, reluctant to yield place to new. In 1767, Ellis, to whom we are indebted for the description of many species of coral, compared them to a beehive, into the cells of which animals make their way by chance and settle there. In 1813, Cavolini considered a coral colony to be a single, many-headed animal. These ideas deviated considerably from Peyssonnel's findings, and it was really Gottfried Ehrenberg who, in 1831, was the first to pronounce a coral colony to be a live aggregation of many generations that remain together in a living, organic union.

In the time that followed, attention was focussed primarily on the skeleton of the coralline animals, and it was described in detail by Edwards, von Koch, Jourdan and Ogilvie. More than 2,000 species of stony corals were described by Linné, Ellis and Solander, Dana, Edwards, Ehrenberg, Klunzinger,

Duerden, Vaughan and many others. With the establishment of a satisfactory preserving technique, von Koch, von Heider, Fowler, Duerden, Matthai and others finally turned to the examination of the soft body, the coenosarc. These studies produced a good deal of detailed morphological information, but for a long time, the living animal was forgotten. It was not until this century that intensive experimental work on corals was begun. And since about the fifties, it has been possible to use newly developed diving techniques to examine corals and coral reefs in their natural biotope. Nevertheless, so many questions still remain unanswered today that no survey of the subject can claim to be comprehensive. Yet this only makes the enigmatic creatures the more fascinating.

Divers in the reef

Whereas a knowledge of the existence and distribution of coral reefs was closely linked to the discovery and exploration of the oceans of the world, scientific investigation of the reefs and of reef-building organisms is bound up with developments in diving technique. The use of ships on research expeditions is impracticable since they are constantly in danger of running aground. Moreover, it would be impossible to use traditional fishing equipment such as trawls, dragnets and grab dredgers, since they would become entangled and torn by the many jutting rocks and spurs. Diving is really the only way to investigate a coral reef.

In 1884, the French zoologist Milne Edwards was probably the first scientist to examine the underwater world, working with piped-air diving apparatus. The number of those who followed him was small at first, because working with heavy equipment made considerable technical demands. The pumping plant had to be worked from a boat or pontoon. The diver is attached to the boat by air-pipes and movement is laborious.

Beebe, an American zoologist, who later reached a depth of almost 1,000 m in a pressurized bathysphere, undertook the first exploratory descents among coral reefs.

In the meantime, three simple pieces of diving equipment that had been known and used for years, now underwent further development. Reed stems such as ancient Greek warriors had used for breathing through while making an underwater attack on an enemy, produced the snorkel; transparent tortoiseshell eye-covers that for centuries had enabled harpooners hunting off tropical shores to see their prey clearly under water, evolved into diving masks; and from the swimming shoes woven out of palm leaves by South Sea Islanders, came the rubber diving flippers that provide such effective momentum to the swimmer that his arms remain free to carry out other tasks.

In the early forties, an Austrian, Hans Hass, was the first to use self-contained oxygen-circulation breathing apparatus to study the Mediterranean reteporids known as "Neptune's veils" (Reteporidae), marine bryozoa or sea-mats of filigree-like delicacy. He introduced diving as a method of biological research. The results obtained in this way were vastly superior to those gained by traditional operating methods, just as results obtained with an electron microscope surpass those with an optical microscope—and of course, especially so among inaccessible coral reefs. Later on, oxygen-circulation breathing apparatus was replaced by the safer compressed-air diving equipment developed by Costeau and Gagnan.

By the fifties, the economic importance of the coral reefs was obvious and intensified exploration inevitably followed. The necessary finances for equipping scientific stations, ships and research teams were made available both by national institutions and private firms. Since then, countless biologists, palaeontologists and geologists have devoted themselves to a study of coral reefs and their inhabitants. Many of them work as divers (Ill. 1,4). They examine the structure, age and growth of the reefs. They carry out measurements and collect specimens to find out about currents, the effects of wave action, temperature and conditions of light, to discover the quantities of oxygen, seasalts, nitrates and phosphates in the water of the reefs. They carry out experiments on live corals, placing a glass jar over them and, after a certain length of time, measuring

the gaseous metabolism (Ill. 5). They introduce a layer of sediment over coral colonies and observe whether, in spite of being sessile organisms, they are nevertheless able to free themselves from loose clay and sand. They isolate organisms living within the colony to discover their degree of dependence upon corals; they study the environmental conditions under which the various animals of the coral reef live.

The effects of man's presence among the reefs are already becoming apparent: poisonous substances are being intro-

duced into the sea, power stations raise the temperature of the water, underwater explosions throw up debris by the ton, sewage pollutes the lagoons. It is urgently necessary to discover not only the potential that exists for the development of the reefs, but also the degree to which the organisms living there can tolerate the wide variety of harmful factors, so that it will be possible to put a stop to destructive exploitation and pollution before more coral reefs fall victim to the rising wave of destruction that threatens them.

Stony corals

their structure, growth and diversity

The name coral is given to animal colonies built up of polyps that can be members of various systematic groups, but that all belong to the phylum Cnidaria, which includes the classes Hydrozoa, Scyphozoa, Cubomedusa and Anthozoa (Fig. 1). Hydrozoa, mostly very delicate and graceful in structure, and Scyphozoa, widely known in the form of large, gellatinous jellyfishes, exhibit a typical alternation of generations with two phases, the polypoid and the medusoid. The polypoid generation is sessile (it grows attached to a surface) and reproduces by budding, forming polyp colonies. Apart from a few exceptions, the medusoid phase arises from the hydroid, and carries male and female generative cells. The larvae develop initially into polyps. In the Cubomedusae, a new medusa develops directly from the larva. On the other hand, the larvae of Anthozoa develop exclusively into polyps (Fig. 2).

Although jellyfishes moving about in the sea seem very different from polyps that grow attached to rocks, both have a basically similar structure: a hollow cylindrical body made up of an outer skin, a supporting lamella and an inner skin, the open, upper end of which is the mouth opening. The mouth is surrounded by a ring of tentacles. The sessile polyp forms have the mouth directed upwards, the jellyfish floats in the water with the mouth underneath and the tentacles hanging down in the water. All of them are armed with stinging cells (cnidoblasts), that have given the phylum its name.

The class Anthozoa contains representatives with polyps measuring some millimetres or decimetres. Many species develop brightly-coloured colonies and when these are covered with radially symmetrical flower-like polyps, they resemble flowering plants, and indeed are sometimes called "flower animals" or "flower polyps". The polyp body is chambered by folds in the wall of the digestive cavity. Depending upon whether the forms possess symmetrical radial arrangement in multiples of six or eight, they are divided into Hexacorallia and Octocorallia. A summary of the two subclasses appears in Fig. 1.

Many of the sessile cnidaria have chitinous, horny or stone-like skeletons. Calcium carbonate is secreted in the form of spicules or sclerites that are stored in the tissue or joined to one another in bark-like bundles. Others grow rigid limestone skeletons. These include the Hydrozoa of the orders Milleporina (Fire Corals) and Stylasterina (Filigree Corals); among the Hexacorallia, the Scleractinia (Stony Corals); and of the Octocorallia, the Stolonifera (Organ Corals) and Coenothecalia (Blue Corals). Among Stony corals there are some hundreds of species.

Structure of the body

A tropical beach is often strewn with irregularly shaped, white rock masses. On closer examination, these are found to be a variety of blocks, fragments, slabs and branches made of limestone. They show an immense number of pores and holes that are criss-crossed inside by fine, radially arranged ribs. These are coral skeletons, broken away from a reef by a stormy sea and thrown up on the beach. The holes once housed many individual coral animals. Anyone wanting to take a few coral skeletons home with him, would do well to take them from here and avoid causing damage by breaking a living colony from a reef. The skeletons are clean and many still exhibit the fine structure of aragonite crystals that have cemented together. They consist of 98 to 99.7 per cent calcium carbonate together with small quantities of magnesium carbonate and traces of organic components.

If a compact lump of coral with quite large coral cups is split open by means of a hammer and chisel, the cups or corallites are found to continue in the form of tubes in a radial arrangement down to the inside of the colony (Fig. 4), interrupted by supporting lateral platforms. A corallite consists of base plate, walls and partitions. There are also additional supporting elements in the form of bars, ridges, rods, crenellations and spicules, which make the structure a complicated one. All of these are specific in form, but frequently vary so greatly as a result of modification by environmental factors, that they are not a reliable guide to species. The outward-pointing spikes,

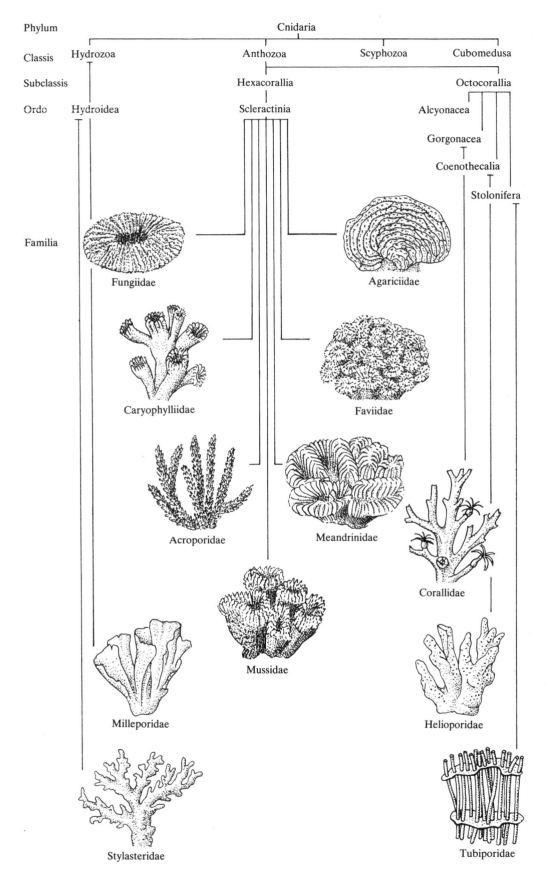

Phylum			Cnidaria			
Classis	Hydrozoa		Anthozoa	Scyphozoa	Cubomedusa	
Subclassis			Hexacorallia		Octocorallia	
Ordo	Hydroidea		Scleractinia		Alcyonacea	
					Gorgonacea	
					Coenothecalia	
					Stolonifera	

Familia

Fungiidae

Caryophylliidae

Acroporidae

Milleporidae

Stylasteridae

Agariciidae

Faviidae

Meandrinidae

Mussidae

Corallidae

Helioporidae

Tubiporidae

1 *Systematic position of Scleractinia (Stony corals) within the phylum Cnidaria (Original)*

thorns and needles make it difficult for other animals to attack the polyps that have retracted into the corallite. The coral cups are embedded in a supporting limestone framework that unites the colony into a cohesive whole.

The polyp is saccular. Simple in structure, it consists of only two layers of tissue, an outer and an inner skin, between which there is a very thin supportive lamella. The muscular system is only slightly developed. It consists of cutaneous cells that extend as fine processes between the two layers of cells. When they narrow, the polyp and tentacles contract. When cells are exhausted, their place is taken by undifferentiated replacement cells that move across from the supportive lamella.

Below the oblong mouth opening is the digestive cavity. The inward folds of the body walls or mesenteries that curve round the septa provide further extension of its active surface; the number of glandular cells is increased, larger quantities of gastric juices are secreted and more food can be ingested. From feeding muscle-cells situated in the endoderm, processes extend to take in food. The mesenteries carry tube-shaped structures known as mesenterial filaments. Two ciliated grooves extend from the mouth into the body cavity. The coordinated pulsation of cilia ensures a continual exchange of fresh water in the stomach. The polyps of a coral colony are linked together by the soft layer of live tissue that extends through the entire colony and by the complicated system of canals that ramify through the corallum. These enable the polyp that has obtained large quantities of food to pass food juices that it produces to other individual polyps in the colony.

A loose network of nerve cells contains sensory cells and ganglion cells. Electron-microscopic examination has shown that impulse-conduction systems and muscle-control mechanisms are similar to those of higher animals.

From larva to coral colony

In the hermaphroditic (bisexual) or heteroicous (unisexual) species of stony coral, the sexual products develop in the testes and ovaries that are situated in the mesenteries. In bisexual species, ripening of the egg- and sperm-cells takes place at different times. This avoids self-fertilization. The sperm cells are expelled through the mouth into the water, make their way by swimming to another polyp, pass through the mouth opening into the digestive cavity and there fertilize the ripe egg cells. These develop into ciliated planula-larvae that make their way out of the parent polyp.

The majority of this host of larvae, numbering millions, is eaten. After days or weeks spent swimming about, a few manage at last to find a clear space on the densely colonized rocky bed. A larva settles and forms a tiny, semispherical polyp body. It immediately begins to lay down a limestone skeleton. By means of a hard basal disc, the young stony corals cement themselves firmly to the rock. Very soon, they have surrounded themselves with a protective wall of limestone, the top of which is armed with sharply pointed prongs and spikes, and within which they live.

Sexual reproduction is followed by asexual propagation. As they continue to grow, most of the colonial stony corals produce buds outside the fringe of tentacles, or fission in the area surrounding the mouth (the peristome) produces two polyps, which divide into four and then into eight polyps. In this way, the variously-shaped coral colonies develop. Transverse division of the polyp is rare in stony corals, occurring only in solitary species. For example, young specimens of Mushroom corals of the species *Fungia* reproduce by splitting off a columnar disc from the main body (Fig. 3).

The colonies that develop are for the most part disc-shaped, covered with a crust, lumpy to semispherical in shape, columnar and delicately or coarsely shrub-like in form, and they often achieve a considerable size. Although Pore corals of the genus *Porites* have tiny polyps measuring only one or two millimetres, there are some species that produce massive colonies 4 m high and 6 m in diameter. A colony of this kind certainly consists of millions of individual animals. It is more than 400 years old. A product of asexual reproduction, it can be thought of as the multiplied substance of that first single young polyp, indeed, one could go further, and because, as they continuously deposit skeletal limestone, the coral animals grow out further and further towards the upper surface, it is possible to imagine the original living polyp under the millions of daughter polyps. In that case, the Pore corals that form very large colonies would have to be counted among those animals that live to a very great age.

Although the individual coral cups (corallites) offer stability and protection to the soft-bodied polyps, the formation of colonies brings with it additional advantages that help to ensure the continuation of the species and of the coral organism. The skeletal elements are joined together in such a way that a degree of rigidity is achieved which, though it varies in degree depending on species and shape, is yet adequate in each case to meet the hydrodynamic con-

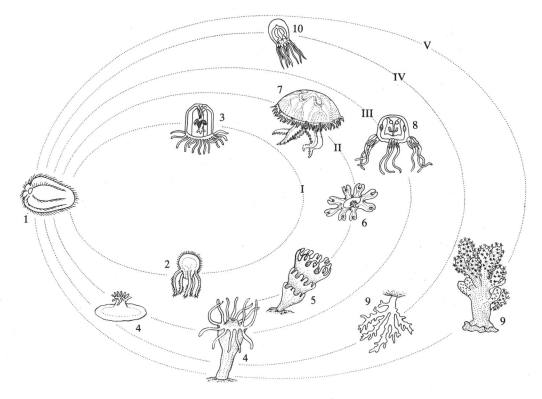

2 *Development cycles of Cnidaria reflect the relationships existing between the different major groups. I. Trachylina, II. Scyphozoa, III. Cubomedusae, IV. Hydrozoa, V. Anthozoa.*
1 planula, 2 medusal larva, 3 trachymedusa, 4 polyp, 5 strobila, 6 separated ephyra, 7 scyphomedusa, 8 cubomedusa, 9 polyp colony, 10 hydromedusa (After Hill & Wells from Moore, 1963, supplemented)

ditions that prevail. The efficient use of calcium carbonate encourages rapid growth of the colony and of the skeletons, so that in a relatively short time, the coral raises itself above the substratum, and many individual polyps are able to escape the effects of the heaviest sedimentation. The larvae expelled by polyps situated at a higher level reach the higher layers of water and have a greater chance of drifting a considerable distance. They are able to establish new colonies away from the parent growth and extend the area of their distribution. Finally, the rapid growth of the colony restricts the field for competitors or even eliminates them completely. By contributing to the extension of the reefs, stony corals in particular constantly enlarge their share of the sea bed and so increase the beneficial effects of this fascinating phenomenon.

The protective armament

Stinging cells (cnidoblasts) are an effective means of defence against many of the enemies of stony corals. The only creatures they cannot harm are those that specialize wholly or partially in a diet of corals. But normally these occur in such small numbers that they pose no real threat to coral stocks.

Stinging cells are also used to catch prey and are situated primarily in the tentacles and the mesenterial filaments. But they may also be distributed over the tissue of the entire outer skin. They are formed by undifferentiated interstitial cells that constantly renew the supply as it is required.

The structure and function of the microscopically small stinging capsules (cnida or nematocysts) contained inside the stinging cells are complicated. Inside the tiny bladder-like vessel holding the poison, there is a delicate long hose like the inverted finger of a glove and tightly rolled. From the cell projects a pointed or whip-like process. This cnidocil works

like the contact detonator of an explosive device. If a quarry touches the process, the stinging capsule "explodes", the filament is eversibly discharged (Ill. 8) and, together with hundreds of filaments from other cells, encircles the quarry, penetrating skin or shield and injecting paralyzing poison into the body (Fig. 5). The action is without influence by the nervous system, occurring directly and at lightning speed within 0.003 to 0.005 seconds. In addition, several tentacles reach out towards the victim, so that it is rapidly dealt an increasing number of venomous injections. If nevertheless it threatens to tear itself free, the long mesenterial filaments are brought into use and extended from the mouth or other parts of the body. They also encircle the prey and subject it to further poisonous secretions. It is not difficult to see how even moderately-sized organisms are unable to escape from this battery of hundreds or thousands of venomous injections. The paralyzed prey is now carried by the tentacles into the mouth of the polyp and drawn into the stomodaeum by the mesenterial filaments.

Although all cnida function on these principles, many different forms exist. Spirocysts and nematocysts are probably the principal ones to occur in stony corals. Inside spirocysts, a fine adhesive thread winds round the central tube. Nematocysts are furnished with bristles or barbs and exhibit considerable functional variations. Their differences are so strongly marked that they have been used as criteria in solving problems of taxonomy.

The evolution of such a complicated cell in the primitively organized Cnidaria indicates the enormous potentialities of form and function that exist within living matter. They illustrate the difficulties and imperfections inherent in any mechanistic classification of natural organisms and in the so-called genealogical trees of the animal kingdom. Certainly the natural origin and evolution of

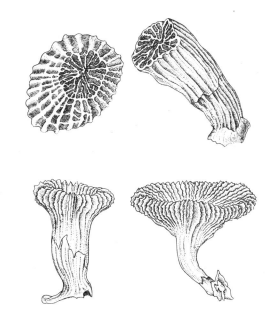

3. Vegetative reproduction in a Mushroom coral (Fungia) *that is initially attached by a stalk (anthocaulus) to the substratum, from which the anthocyathus subsequently detaches itself and grows into a solitary coral lying freely on the ground. (After Bourne from Pax, 1925)*

living beings has been confirmed a thousandfold since Darwin, and is today an element in man's conception of the world, but nevertheless we are, in many cases, still far from understanding the natural relationships that exist between groups of organisms.

Thousands of species inhabit the oceans

Approximately 2,500 living and 5,000 extinct species that make up the order Scleractinia are grouped together in a system based on similarities in skeletal characteristics. Natural relationships have not been taken into account. Nor can we look for clarification in the near future, since results of morphological examination of the soft bodies of stony corals have so far proved inconclusive. Moreover, research to this end in the spheres of electrophoresis, immunology and chromosomal structure by mathematical methods and by analysis of variation have also proved unsuccessful. So the existing classification of the Scleractinia can merely provide a clearer overall picture.

Here we can present only a few typical corals that are particularly common in reefs (Ill. 9–23). But in order to determine family and genus accurately, more characteristic features are necessary. An additional difficulty is that there is great variation in coral form within species, depending on differing environmental conditions, and therefore characteristics are highly variable. So determination of species is a subject for the specialist. Popular names for corals are rare in our language. So a series of new vernacular names has been added to those already familiar.

Family Thamnasteriidae. Colony encrusting, irregular, rarely disc-shaped, not branching. The corallites are very small and embedded superficially in the skeleton of the colony; they have the appearance of radially symmetrical blossoms, the central rod or columella resembling stamens and the septa radiating to the outer wall, petals. Genus *Psammocora* (Blossom Star coral), Indo-Pacific.

Family Pocilloporidae. Colony medium-sized, branching. Corallites are minute and occur in large numbers, open, with short septa and columella. Genera *Stylophora* (Styloid coral), Indo-Pacific, common in shallow, calm water; *Madracis* (Tessellated coral), circumequatorial; *Pocillopora* (Claviform coral), Indo-Pacific; *Seriatopora* (Needle coral), Indo-Pacific, also on sandy ground.

Family Acroporidae. Shape of colonies varies. Corallites small to minute, columella absent. Genera *Astreopora* (Porous Star coral), Indo-Pacific, on gravel-covered or sandy substrate; *Acropora* (Staghorn coral, Umbrella coral), more than 200 species, circumequatorial, frequently massive, principal reef-forming species; *Montipora* (Micropore coral), Indo-Pacific.

Family Agariciidae. Colonies are foliaceous and disc-shaped, rather rarely massive. Corallites in rows, more or less markedly countersunk in depressions. Genera *Agaricia* (Fluted coral), Atlantic; *Gardineroseris* (Honeycomb coral), Indo-Pacific; *Pavona* (Ribbed coral, Starry columnar coral), Indo-Pacific; *Pachyseris* (Gramophone-record coral), Indo-Pacific, in deep-water region; *Leptoseris* (Undulating coral), circumequatorial, in deep-water region.

Family Siderastreidae. Colonies show varied growth forms, corallites fairly small. Genera *Siderastrea* (Funnel coral), Atlantic; *Coscinarea* (Torose coral), Indo-Pacific.

Family Fungiidae. Discoidal to elongate-oval solitary corals or coral colonies, remarkable for numerous and large septa. Indo-Pacific. Genera *Fungia* (Fungus or Mushroom coral), *Cycloseris* (Small Mushroom coral), *Ctenactis* and *Herpolitha* (Tongue corals), *Halomitra* and *Parahalomitra* (Hat corals), all lie

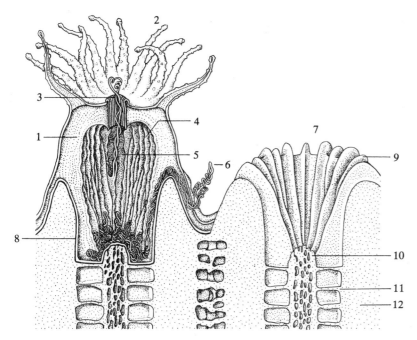

4 *Structural diagram of a stony coral colony.*
1 polyp, 2 tentacles, 3 mouth opening, 4 mesenteries, 5 gastrovascular cavity, 6 mesenterial filaments, 7 calyx, 8 theca, 9 septum, 10 columella, 11 dissepiments, 12 coenosteum (Original)

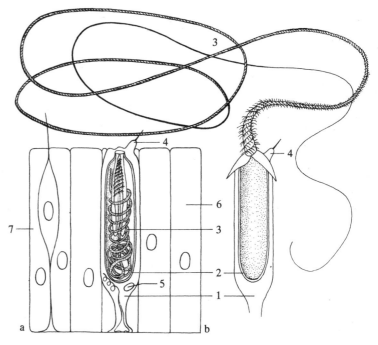

5 *Stinging capsules (nematocysts, cnida) with thread a) rolled and b) extruded. 1 cnidoblast (stinging cell), 2 nematocyst, 3 coiled tubular filament, 4 cnidocil, 5 cell nucleus, 6 epidermic cell, 7 sensory cell (Original)*

free, unattached to a substratum, *Podabacia* (Plate coral), grows in deep water.

Family Poritidae. Colonies variously shaped. Corallites small to minute, closely united, with columella and trabeculae. Genera *Porites* (Pore coral), some 50 species, circumequatorial, many form coral pillars more than 8 m high, important in reef building; *Goniopora* (Coarse Pore coral), Indo-Pacific; *Alveopora* (Sponge coral), Indo-Pacific.

Family Faviidae. Colonies vary greatly in form. Corallites are solitary or fused together and meandroid. Genera *Favia* (Star coral), Indo-Pacific; *Favites* (Angular Star coral), Indo-Pacific; *Echinopora* (Rough Star coral), Indo-Pacific; *Montastrea* (Button coral), circumequatorial, important reef-builder in oceans of Central America; *Cyphastrea* (Crinkly coral), Indo-Pacific; *Diploastrea* (Flower Star coral), Indo-Pacific; *Goniastrea* (Garland coral), Indo-Pacific; *Diploria* and *Colpophyllia* (Fine and Coarse Atlantic Brain coral); *Platygyra, Leptoria* and *Oulophyllia* (Indo-Pacific Brain corals); *Trachyphyllia* (Conical Brain coral), Indo-Pacific, in sand; *Manicina* (Plicate coral), Atlantic, in reefs and on sandy ground; *Hydnophora* (Pimpled coral), Indo-Pacific; *Cladocora* (Small Tube or Lawn coral), Atlantic and Mediterranean, also on sandy ground.

Family Rhizangiidae. Stump corals. Colonies with small corallites resembling short stumps that have the appearance of being solitary but which are linked together at the base by a thin layer of skeleton and tissue. Includes genera such as *Astrangia, Culicia* and *Phyllangia.* In reefs, they are always in dark areas.

Family Oculinidae. Colonies branching, encrusting, cushion-shaped or columnar. Genera *Oculina* (Eye coral), Atlantic; *Acrhelia* (Thorn-bush coral), Pacific; *Galaxea* (Scalpel coral), Indo-Pacific. The genera *Sclerhelia, Bathelia, Madrepora,* that are also branching forms, and some others live at depths of 100 m to 1,000 m or more.

Family Meandriniidae. Colonies massive or pillar-shaped, corallites meandroid or solitary. Atlantic. Genera *Meandrina* (Meandrine coral); *Dendrogyra* (Candelabra coral); *Dichocoenia* (Tuft coral).

Family Merulinidae. Colonies are leaf-shaped or irregular, corallites linked forming "valleys". Indo-Pacific. Genera *Merulina* (Ribbed Leaf coral); *Scapophyllia* (Irregular Ribbed coral).

Family Mussidae. Solitary or colonial corals with large to very large corallites, the margins of the septa are strongly dentate or thorny. Genera *Lobophyllia* (Indo-Pacific Thick-stemmed Umbellate coral); *Mussa* (Atlantic Thick-stemmed Umbellate coral); *Scolymia* (Atlantic Large Star coral), *Parascolymia* (Pacific Large Star coral); *Isophyllia* (Atlantic Plicate coral); *Symphyllia* (Indo-Pacific Plicate coral); *Mycetophyllia* (Flat Plicate coral), Caribbean; *Isophyllastrea* (Atlantic Cactus coral); *Acanthastrea* (Indo-Pacific Cactus coral).

Family Pectiniidae. Leaf- or disc-shaped, colonies thin and delicate. Indo-Pacific. Genera *Echinophyllia, Oxypora, Mycedium* and *Physohyllia* (Large Leaf coral); *Pectinia* (Crumpled-paper coral).

Family Caryophylliidae. Many genera and species, predominantly solitary. Most are ahermatypic at depths between 100 and 4,500 m. Also a few colony-forming species growing in reefs, with compact or branching growth. Genera *Lophelia* (Deep-sea Shrub coral), in all oceans at 60–2,000 m; *Eusmilia* (Small Bouquet coral), Atlantic; *Euphyllia* (Large Bouquet coral), Indo-Pacific; *Plerogyra* (Bladder coral), Indo-Pacific; *Gyrosmilia* and *Physogyra* (Flat meandrine corals), Indian Ocean.

Family Dendrophylliidae. Some colonial and many solitary species, mostly deep-sea forms, mostly ahermatypic. In reefs, predominantly the colony-forming orange-red *Tubastrea* (Tube coral) and dark-coloured *Coenopsammia and Den-*

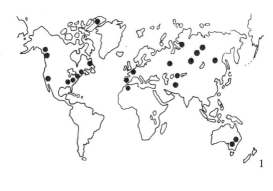

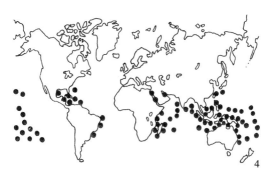

6 *Displacement of the tropical zone as a result of shifting of the equatorial belt of reefs in the course of geological history. 1 Cambrian, 2 Devonian, 3 Cretaceous, 4 Recent (After Schwarzbach, 1950)*

drophyllia (Tree corals), in all oceans from 0 m to 1,500 m, and the Indo-Pacific *Turbinaria* (Foil coral).

Since almost all corals keep their tentacles contracted during the day, the skeleton is usually clearly visible under water. So it is not necessary to break off sections of corals in order to make a preliminary general identification. To examine small corallites, a folding magnifying lens worn on the wrist when diving is a useful aid. Approximate identification at the rock face makes it considerably easier subsequently to name the corals that have been photographed by underwater camera, although the unequivocal establishment of genus will often prove difficult. It might be some consolation to know that of all identifications carried out by zoologists, that of stony corals is one of the most difficult, and in many cases, even specialists have difficulty in distinguishing between species.

Conquest in four dimensions

On this earth, which may seem to us the embodiment of all that is solid and firm, the distribution of land and water masses has undergone fundamental changes in the course of its existence. During processes of condensation and cooling that continued for millions of years, a large landmass came into existence, Pangaea, that was washed round by Panthalassa, the primitive Pacific. It was probably not until the Mesozoic that Pangaea split into several smaller continents. After that time, the Tethys, a broad sea stretching across the world from east to west, linked primitive Pacific with primitive Pacific. Its waters provided the conditions necessary for the development of living matter.

Very distant relations of the corals, the protomedusae, were already living in the Pre-Cambrian era. Following an early

glacial period, huge quantities of melt water caused flooding of the shelf regions. The formation of this shallow water area may have occasioned the development of benthic organisms. And these could well have produced skeleton-building, coral-like animals. Fossil finds, however, date back only some 440 million years, to the start of the Ordovician period. They belong to two great orders, Rugosa and Tabulata. Features they possessed in common with the stony corals of today are the typical coelenterate structure and the secretion of a hard skeleton. They also occurred both in colonies and as solitary individuals. Cambrian reefs built by them, that have become part of the mainland, are found predominantly north of the Tropic of Cancer, in North America, Europe and Asia and even within the Arctic circle in Greenland (Fig. 6).

Not only was the distribution of sea and land at that time different from that of today, but in addition, the equatorial, tropical zone extended much farther northwards and only gradually moved southwards to take up its present position. Although reef formations of this kind were built by different coral-like colonies, it is reasonable to assume warm sea temperatures for their development (Fig. 7). The ancient corals lived for about 200 million years until the end of the Permian. They may have died out as a consequence of geological and hydrographical disturbances linked with the development of the Tethys. After that, there seem to have been no corals living on the earth for some millions of years. The reefs of the Paleozoic had been formed by various sessile organisms, sponge-like archaeocytes and stromatopores, rugose and tabulate corals and moss animalcules or bryozoa. Apart from bryozoa, all have become extinct.

About 180 million years ago, in the middle of the Triassic period, the newly evolved Scleractinia, the stony corals of today, set about continuing the work of

constructing underwater mountain ranges. They had probably developed polyphyletically from several actinia-related ancestors. The early stony corals belonged to a series of different families, of which the Thamnasteriidae and Astrocoeniidae still exist today. Quite small coral banks and reefs built at that time existed in what is today the south and south-east of Europe, and in south and south-east Asia, in elevated land masses. They grew up in the Tethys. Stony corals achieved their widest distribution in the late Trias, between latitudes 60° N and 10° S.

In the early Jurassic period, certain groups of corals died out. But about 150 million years ago, the Caryophylliidae appeared, with their wide variety of species, and with them the first solitary corals. This proved to be an important turning point, in terms of quality, in the further development of the Scleractinia, for now the number of species increased greatly. In the Upper Jurassic, the Faviidae were added, of which many representatives still occur today. The corals formed extensive reefs that became an important feature of the Tethys and its secondary seas. The most northerly reefs from this period have been found in England, Central Europe and Japan, the most southerly in East Africa. The reef belt lay some 20° further north than it does today.

As the equatorial tropical belt was displaced progressively southwards, the coral reefs of the Cretaceous period appeared in Southern Europe, from France to the Crimea, in North Africa, India, Venezuela, Mexico and Texas. Once again, representatives of newly-evolved families that still exist today, such as Siderastreidae, Rhizangiidae and Micrabaciidae, took the place of others. The number of species of solitary corals continued to increase. They became established in all zones of the Tethys from the coast to the deeper parts of the ocean. With the appearance of the Pocilloporidae, Poritidae and Dendrophyli-idae towards the end of the Cretaceous, about 140 million years ago, the stony coral fauna had achieved fundamentally the form it has today.

Because of fluctuations in prevailing conditions in the Tertiary, only a few reefs developed on the edges of the Tethys. In them, the predominant coral groups were almost the same as those of today. Events of far-reaching importance were the virtual extinction of "European" corals, and the splitting of the Tethys. The final separation of the Mediterranean from the Indian Ocean was completed. As a result, the potential dispersal route for the eastward spread of the organisms from the Central Tethys was closed. When the Central American isthmus arose, the western part of the Tethys was also cut off and the Atlantic came into being. At the same time, the two great reef regions of today in the Atlantic and Indo-Pacific were formed. The Pocilloporidae, Acroporidae and Poritidae developed. Now reefs occurred only up to 35° North, which represents the approximate extent of their dispersal today.

The formation of recent coral reefs was influenced predominantly by glacial and post-glacial conditions. The Atlantic was open to the North Pole, but closed off from the Indo-Pacific. The Rocky Mountains that extend from north to south were no obstacle to the southerly advance of the continental ice. Overall, the effects of glaciation were much more marked here than in the reef areas of the Indo-Pacific that were protected from the continental ice by the mighty barriers of mountain ranges running east-west from the Pyrenees, by way of the Himalayas to the Aleutians. But the inland ice masses held vast quantities of water bound as ice, and the level of the oceans sank by 70 to 200 m. As a consequence, many reefs throughout the world died. In addition, in parts of the Atlantic that had cooled considerably, many species of coral became extinct; these included *Fungia, Pocillopora, Stylophora, Montipora, Alveopora, Hydnophora, Galaxea* as well as Organpipe coral *(Tubipora)*

and Blue coral *(Heliopora)*, all of which have been found in fossil reefs of the West Indies. However, more robust species survived the last glacial epoch in places of refuge farther south. After the gradual post-glacial warming of the Central American oceans that took place over some thousands of years, they once again took possession of the Caribbean. These greatly reduced coral stocks, in many ways resembling a relict fauna, nevertheless proved vital enough to restart the work of reef building.

Since stony corals first appeared in the Triassic, they have undergone a very successful evolution, in the course of which they have conquered the entire biotope from deep sea to coastal waters. In doing so, they divided into a multiplicity of species, of which the later-developing ahermatypic (non-reefbuilding) species preferred the cold and dark regions of the ocean, and the hermatypic (reef-building), predominantly colonial species, the coastal regions. In their ability to create reefs, hermatypic corals have exhibited great biological forces.

7 Reconstruction of a Devonian reef in present-day Poland with 1 massive Stromatopores, 2 branching and 3 massive Tetracorals, 4 massive Tabulata, 5 Sea Lilies, 6 Auger snails and 7 shell-bearing Cephalopods (After Rozkovska, 1980, modified)

Environmental factors

as the basis of existence for stony corals

For many millions of years, the earth has provided living matter with conditions that have led to the development of variously constructed groups of organisms and different forms of life. The extent and nature of dependence of the individual systematic units upon their environment vary greatly. Many poikilothermic groups of animals—those with variable blood temperature—such as threadworms and ringed worms, crustaceans, beetles and bugs, have representatives both in water and on land, and are found from the Arctic to the tropics. Not so the Scleractinia. Limited exclusively to the sea, their distribution potentialities are restricted. Their sessile mode of life confines them to the substratum. In addition, hermatypic corals occur only in the shallow waters of tropical oceans. Nevertheless, corals as a whole are an extraordinarily successful group of animals, particularly remarkable for their ability to build vast reefs. It is interesting to trace some of the factors that have enabled them to achieve this success.

The menu

"Blue is the colour of the desert regions of the ocean", said Wüst, a wellknown oceanographer. His words aptly depict certain ocean regions that contain only meagre quantities of suspended matter, and where, as a result, transparency of the water is high. Ocean "wastelands" of this kind, lacking in living organisms, are common in the tropics, because there, upwelling currents bringing water from great depths, enriched with mineral nutrients in solution, occur only rarely. Yet it is precisely here that coral reefs thrive. How is this possible?

As far as finding food is concerned, stony corals are true masters of the art of economy. They may be said to surpass even those best of cooks who are able to conjure up good meals out of very little. If they lack one ingredient, they use another. So far, we know of five ways in which reef-building corals feed.

The first is by catching small animals that form part of the drifting zooplankton. The size of the prey organisms depends upon the diameter of the polyps and the length of the tentacles of the individual species. The small-polyped Staghorn, Styloid and Pore corals trap only extremely small plankton measured in microns (μ), predominantly larvae, protozoa and rotifers. Species with large polyps, such as Brain or Star corals take planktonic crustaceans or small worms. The giant polyps of Mushroom and Umbellate corals even consume marine annelids, crayfish and small fishes. Because coral is a sessile organism, the supply of food is necessarily a matter of chance. But corals are selective. In experiments, diatomaceous algae were rejected and pieces of crab accepted. When mixed food was presented, vegetable components were expelled. Undigested algae were found in the stomach of corals. So stony corals could be called carnivorous. But that is only half the truth.

The second method of feeding consists of taking in coarse fragments of dead organisms as food. In a coral reef brimming with life, many animals of prey are consumed every day by predators. Small fragments drift down through the water and may land on a coral. During the day, the polyp tentacles are retracted, but the outer surface of the polyp is covered by microscopically fine cilia that are in constant, directed motion, but which are capable of altering the direction of pulsation. When a fragment of edible matter comes in contact with the cilia, they direct it towards the mouth. If it is not to the polyp's taste, the material is whirled away in the opposite direction (Fig. 8).

The third means of obtaining nutrition is concerned with microscopically small food particles. For example, when a predator attacks prey, minute fragments of muscle, tissue and blood cells or products of reproduction, faeces and other materials fall upon a coral colony. In observations carried out on Plicate coral (Manicina areolata), it was found that when blood is presented, the tentacles remain retracted, but immediately after contact, glandular cells situated in the ectoderm are activated by the mechanical and chemical stimulus to secrete quantities of glutinous mucus. This envelops the particle and the pulsation of the cilia conveys it on a network of slime to the mouth. In this way, even bacteria that drift in great quantities over the coral reefs, are utilized as a source of nourishment.

These three methods of feeding are undoubtedly employed by the individual species of coral with a higher degree of differentiation than is generally realized. For example, in the Caribbean, the small-polyped Pore coral and Tessellated coral catch only minute planktonic organisms with their tentacles. Grooved corals make use of mucous webs to gather fine particles of food. Staghorn, Funnel, Star and Meandrine corals use a combination of the two methods.

A fourth means of taking in food seems at first a curious one, but it is undoubtedly used by more marine animals than we might suppose. Useful organic solutions are absorbed through the surface of the body by osmosis. They include glycins, alanins and leucins. Electron-microscopic examination has shown the ectoderm of Scleractinia to be semi-permeable; the path taken by leucin and glycin has been marked by radioactive isotopes and tracked through the ectoderm; in this way, absorption of materials in solution through the surface of the body has been proved. It is possible that these processes are linked to a high-alkaline phosphorus-monosterasis that functions as phosphorus-transferasis to provide the energy for metabolic processes in the ectoderm. The processes are assisted by a morphological peculiarity of the ectoderm. The upper surface of the polyp skin examined by electron-microscopic enlargement is found to resemble the hide of a mammal, with an under-layer and covering-layer of hair. So the coral polyp is furnished not only with a covering of cilia but also among these there are considerably smaller, closely-packed finger-shaped extrusions, the microvilli. They effect a considerable extension of the area of active membrane and increase the effect of permeation. Although the quantity of dissolved nutrients in the water is minimal, their assimilation can be of perceptible benefit. In addition, consumption of energy is less than with the intake of solid food

particles that must first undergo the process of digestion. And so, even in the absence of solid food, starvation is virtually excluded; for stony corals, living in the sea is like living in a very dilute nutrient solution.

The fifth mode of feeding takes place at a high level of ecophysiological development and occurs predominantly in hermatypic stony corals—an example of the fact that complicated biochemical processes operate even in simple organisms and independent of the level of development of the organic structure. It concerns the symbiotic relationship between scleractines and zooxanthellae (Fig. 8) that is of fundamental importance to the development of coral reefs, and provides the key to the understanding of this complex ecological system. Zooxanthellae are tiny unicellular algae, measuring only 11 μ, of the species *Symbiodinium (= Gymnodinium) microadriaticum*. They are flagellate algae which, in this vegetative stage have lost the flagellae characteristic of the group, in which there are many species, and as a result also the ability to swim—doubtless an adaptation to their endosymbiotic mode of life. They exist in large numbers, embedded in the tissue of reef-building stony corals and also in Fire corals, Organpipe, Gorgoneans and Blue corals, in Mangrove jellyfish, in many flatworms, sponges, bivalves, even in single-cell shell-bearing foraminifers and radiolaria in tropical seas. Some million zooxanthellae have been found in 1 cm^2 of endoderm of stony corals. 7,400 such algae have been counted in microscopically small coral larvae. They are taken over directly from the parent in order to ensure the vitally important symbiotic relationship. Brownish-green in colour, it is largely they that determine the colour of the coral reef, where browns, greens and yellows predominate.

Since any symbiotic relationship is mutually beneficial, what advantages accrue to the zooxanthellae and what to

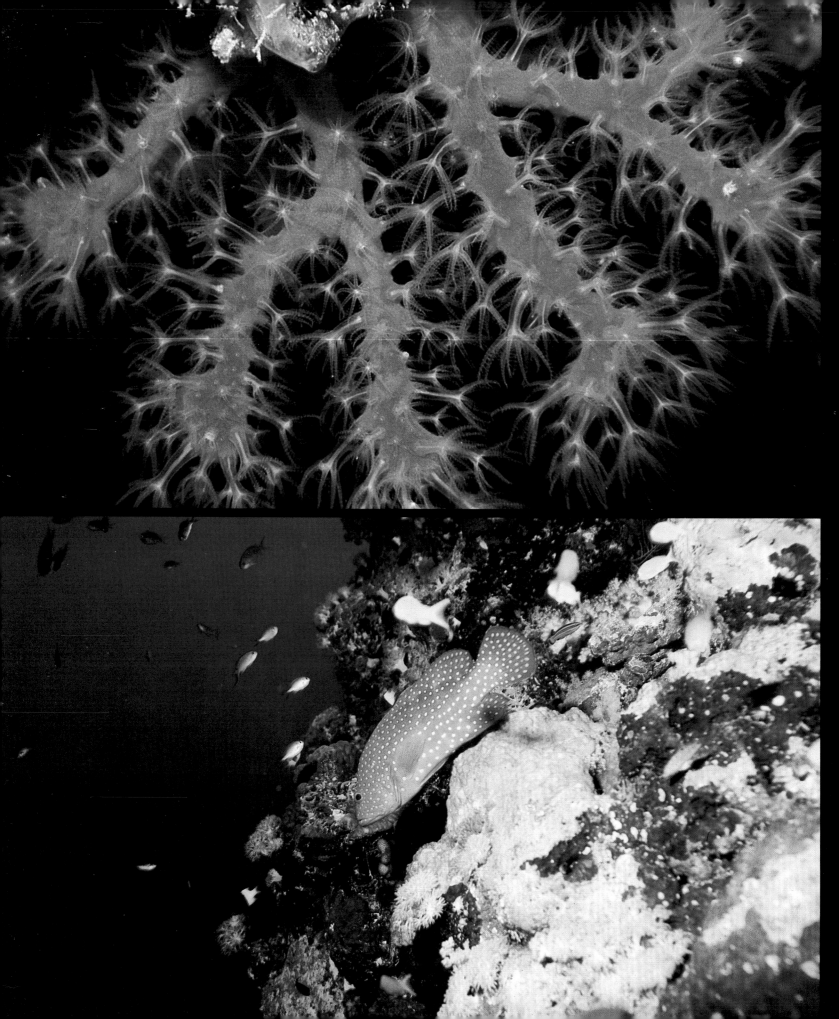

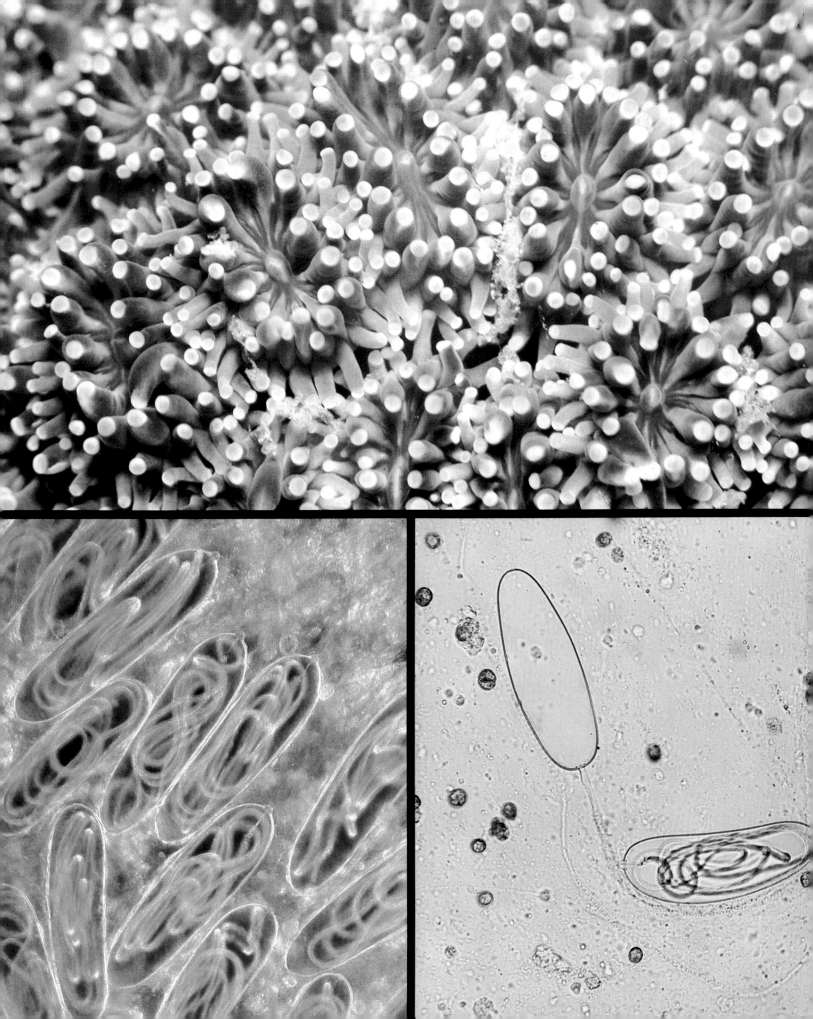

6 *Scalpula coral (Galaxea fascicularis) has tube-like tentacles, blunt and often thickened at the ends. (Red Sea)*

7 *Microphotograph of the tissue of a discus anemone (Actinodiscus). It belongs to Corallimorpharia which have close relations to Scleractinia. Stinging capsules standing close together form an effective defence and killing installation as an arms arsenal. Magnification 420 : 1*

8 *Two stinging capsules, one with a rolled-up tubular thread inside and clearly visible hair-like trigger, the cnidocil, the other with everted penetrant thread. The roundish, yellow objects are zooxanthellae. Magnification 120 : 1*

9 *Orange-red Tube coral (Tubastrea aurea) is one of the few species occurring in reefs that do not have symbiotic association with zooxanthellae and that live in the dark. (Virgin Islands, Western Atlantic)*

10 *Massive Elkhorn corals (Acropora palmata) have short, sturdy stems that branch into thick boughs. (Cuba, Caribbean Sea)*

11 *Bush-like growth of Staghorn coral (Acropora humilis). (Great Barrier Reef, Australia)*

12 *Colonies of Staghorn corals, in this case Acropora pharaonis, that spread in an extended umbrella shape are not unusual in calm water. (Red Sea)*

13 *In calm water, the intricately branching twigs of Needle coral (Seriatopora hystrix) grow into fragile miniature bushes in which small Cardinalfishes (Paramia quinquelineata) seek shelter. (Great Barrier Reef, Australia)*

the corals? The zooxanthellae, living within a rigid spiny skeleton, surrounded by stinging cells, gain protection in the same way as the polyps. Most important of all, they continually make use of the coral's products of metabolism: carbon dioxide produced by respiration as well as excretory products discharged as nitrogen and phosphate compounds. The latter are scarce commodities in the ocean, but are essential to the algae. By means of their chloroplasts (chlorophyll-bearing plastids) in conjunction with light, carbon dioxide and water, they produce a compound and give off oxygen.

$$6CO_2 + 12H_2O \xrightarrow[\text{chloroplasts}]{\text{sunlight}} 6(CH_2O) + 6H_2O + 6O_2 + 674\,kcal$$

The oxygen is used by the coral in respiration. From this remarkable chemical micro-factory, it also receives a series of other nutrient materials such as various sugars and amino acids. Finally, by assimilating the products of metabolism, the algae effect the detoxication of the coral organism—a frequently underrated but extremely important, life-supporting process. By means of these coordinated, mutually complementary physiological processes that take place within an extremely confined space, losses of material and energy are largely avoided, the partners obtain rare but important trace elements and compounds, and by constant feedback and utilization, waste products are kept extremely slight.

The importance of this symbiosis to the coral reef lies in the role it plays in accelerating the process of skeleton-building. The basic materials, calcium and carbon dioxide, are present in sea water in large quantities. How calcium ions are extracted from sea water is not yet known, but they are concentrated at a matrix consisting of mucopolysaccharides (Fig. 9). They combine with the bicarbonate ions produced by metabo-

lism to form calcium bicarbonate (a) and by means of a further reaction, calcium carbonate (b). In accordance with the law of mass action, a large proportion is again dissolved.

a) $CA^{++} + 2HCO_3^- \rightarrow Ca(HCO_3)_2$
b) $CA(HCO_3)_2 \rightarrow CaCO_3 + H_2CO_3$

If, however, H_2CO_3 is split into CO_2 and H_2O by anhydration, and at the same time CO_2 produced in relatively large quantities by the corals is taken up by the assimilating zooxanthellae, the splitting of H_2CO_3 is accelerated. In such a case, the rate at which calcium carbonate is formed is ten times higher than it is for algae-free corals living in deep water and in cold seas. Under the influence of zooxanthellae, hermatypic colonies undergo an increase in size and can take on gigantic proportions. In other groups of animals as well, individuals have been able to develop to a particularly great size under appropriate conditions. For example, the giant foraminifera *Marginopora vertebralis* and the clam *Tridacna gigas* vastly exceed the normal limits of growth.

The process of calcification is also supported in the following way. Calcium carbonate occurs in two crystalline forms, as aragonite and as calcite. Scleractines construct their skeleton of aragonite crystals, together with minimal quantities of magnesium carbonate and organic materials. But orthophosphates and organic phosphates excreted by the animals have an inhibitory effect on the formation of aragonite crystals. At this point, the zooxanthellae intervene again; they divest the host of the inhibitory phosphates and use them in their own metabolism. Both they and the host benefit, and the production of aragonite can continue unimpeded. Like all chemical processes, this too depends upon temperature. The optimum lies between 25°C and 31°C—precisely the temperature range that prevails in most of the reefs.

It is, moreover, striking that only the corals living in symbiotic association with zooxanthellae grow on those parts of the reef freely exposed to the light of the sun, while the parasymbiotic species thrive only in shade and darkness. Clearly the unicellular symbiotic algae and their chloroplasts provide the sensitive coral tissue with effective protection from ultraviolet light. The host corals are thus enabled to settle on a substratum fully exposed to the sun, particularly since even their larvae are equipped with zooxanthellae (see p. 24). An excellent example of a mutually beneficial working system. On the one hand, the light requirements of the zooxanthellae are fulfilled. On the other, it is only here with the support of the symbiotic algae that corals achieve a sufficiently high speed of growth to enable them to compete successfully with other sessile organisms and to construct their own world of the reefs.

In spite of this apparent perfection, there is an associated phenomenon that at first sight seems unclear. Because of the dependence of the zooxanthellae on light, the greatest production of calcium carbonate might be expected to occur immediately beneath the surface of the water. But it takes place lower down, at a depth of 2 to 25 m. The reason for this is that the light intensity at the surface of the water is too great for the symbiotic algae, and in addition, the extent of warming of the water is excessive, since it frequently reaches over 31 °C, and is subject to considerable fluctuations.

Attempts have been made by electron-microscopic observation to follow in detail the process of calcium synthesis. Between the polyp and the skeleton, the cells of the ectoderm deposit extremely fine chitinous filaments, about one hundred-thousandth of a millimetre thick, consisting of a mucopolysaccharide. Since in this area there is a gel-like solution heavily saturated with calcium ions, calcium crystals are precipitated. The chitinous filaments provide the nucleus of crystallization and at the same time determine the directional tendency of the aragonite crystals that accumulate round them, and of the skeletal components that develop from them. In this way, the skeletal structures formed are species-characteristic. It is not difficult to imagine that the environmental factor of water movement could alter the orientation and concentration of the chitin filaments and, by modifying the skeleton, lead to those ecological growth forms that have introduced so much confusion into the taxonomy of stony corals.

The distribution of zooxanthellae in a coral colony is far from uniform. Bladder coral, *Plerogyra sinuosa,* carries only a few symbiotic algae in its tentacles, since in any case, it extends its tentacles to catch plankton only at night. But it has supplementary bladder-like organs, (hence its name), in which zooxanthellae live in large numbers. In the pale light of morning, these vesiculate processes are widely extended, so that the symbiotic algae are exposed to the meagre light to the greatest possible extent. At midday, when the light has become too intense for the algae, the polyp retracts the bladder-like processes perceptibly. But at night, when in any case darkness excludes the possibility of photosynthesis, they are retracted and make room for the tentacles (Ill. 21).

The greatest amount of calcium carbonate can be expected where the largest numbers of *Symbiodinium microadriaticum* are concentrated. In one species of Staghorn coral, the number of zooxanthellae found in the tips of the branches was only a third of that found 6 to 9 cm lower. Yet results showed that the rate of calcium production at the tips was 30 per cent higher than lower down. After some searching, the reason was found. It is quite true that at those parts of the coral where zooxanthellae occur most densely, a surplus of products that are useful in skeleton construction is developed, but they are then carried through connecting channels that permeate the coral tissue to the tips. The consequence is a particularly rapid growth of skeleton at this point.

Examination of coral communities at different depths shows that the number of zooxanthellae is at its highest at a depth of 10–20 m, while at 3–5 m it is only some 40 per cent of this figure. The reason for this is not known. At the same time, the chloroplasts involved in the photosynthetic processes of the zooxanthellae vary depending upon the depth and the differing quality and quantity of light. This could, at least to some extent, compensate for physiological differences arising from the fluctuating number of zooxanthellae (see p. 38).

From what has been said, it is clear that only stony corals living in symbiotic association with zooxanthellae are capable of reefbuilding, and that without symbiosis, biogenic reefbuilding is inconceivable. Consequently this symbiosis

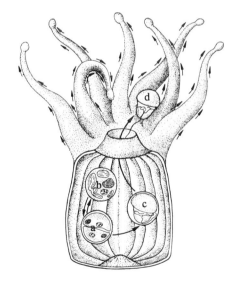

8 Food intake: in many corals, the ciliary currents carry away from the mouth. But if a tentacle is curved inwards, they direct it towards the mouth and food particles can be taken in. Diagram showing different stages in the life cycle of Symbiodinium microadriaticum *inside the coral polyp. a vegetative division, b spherical alga, c a motile stage develops in a host cell, d motile flagellate stage (Original)*

justified the division of the Scleractinia into the ecophysiological groups of reef-building and non-reefbuilding corals. At the same time, the dependence of the symbiotic partner on light explains why hermatypic corals occur most frequently up to depths of 50 m, after that increasingly sporadically, and at depths up to 100 m rarely. Since both symbiosis and increased precipitation of calcium depend on warm water, reef building is restricted to tropical seas. At the same time this provides a plausible biological explanation for the fact that stony corals do not accept as food any chromatophore-bearing plants; to do so would endanger symbiosis with the zooxanthellae. Yet Scleractinia must also have available to them algae-digesting enzymes for use in emergencies, for it has been found that if experimentally deprived of food, they will eat the symbiotic algae, even though plant food is normally rejected. Here too, hunger is the best sauce.

Clearly the utilization of the available nutrient substances and food resources by hermatypic corals is very varied and comprehensive. From an ecological point of view, they behave first as primary producers because, by way of the zooxanthellae, they make use of inorganic substances transformed into organic, then as secondary producers, because as consumers of plankton and foodstuffs in suspension, they feed on living organisms, and thirdly as tertiary producers, consuming detritus and nutrients in solution. The importance of the heterotrophic modes of feeding which are widely estimated to make up 10–20 per cent of the total, is considerably less than that of "autotrophic" feeding carried out with the help of symbiotic algae. Both coral and alga are, of course, able to maintain themselves in their environment more successfully together than separately, and only in partnership, by means of the intensified exchange of constituent materials unique to them, are they able to

create the extensive world of the coral reef and to continue the process of reproduction, even in the middle of the bluest of oceans, in the "desert regions" of the sea.

There are also many predominantly solitary stony corals living outside the reefs, most of them without algal symbiosis. Accordingly, stony corals can be divided into four ecophysiological groups: 1. hermatypic-symbiotic corals that are by far the most common ones in reefs, 2. hermatypic-aposymbiotic corals found somewhat rarely in reefs, 3. ahermatypic-symbiotic corals occurring in the photic zone outside reefs, and 4. mostly small ahermatypic-aposymbiotic corals that are not rare in the dark, cold depths of the sea or in other habitats.

A number of aposymbiotic species, such as the hermatypic *Coenopsammia* and *Dendrophyllia* (Tree corals) or the ahermatypic *Lophelia* (Deep-sea Shrub coral) build sizeable branching colonies which they are probably able to construct only by means of specialized metabolic processes, of which little is yet known to us in detail.

Flourishing growth

A single coral polyp grows rapidly, but large colonies can be decades or even centuries old. In the building of coral reefs, the rate of calcium production, which determines the speed of growth of the coral, must be higher than its rate of destruction as a result of abrasion. From this point of view, the growth processes of stony corals are highly significant.

In order to obtain information on reproduction in the individual polyp and on growth of skeletal material, counts were made, measurements taken and weights checked both at the reef and in the laboratory. In experiments, the quantity of ^{45}Ca und ^{14}C absorbed into the skeletal mass was correlated with the oxygen content of the live tissue. This ratio altered during the growth of the

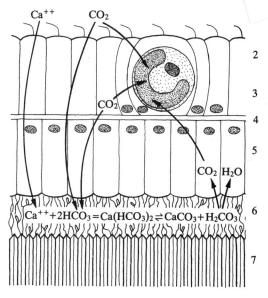

9 Diagram showing development of calcium carbonate in stony corals 1 sea water in gastrovascular cavity, 2 endoderm, 3 zooxanthellae, 4 supportive lamella, 5 ectoderm, 6 matrix with minute crystals of calcium carbonate, 7 limestone skeleton (After Yonge, 1963, modified)

coral colony in favour of the mineral component, that is, the skeleton increases in weight more rapidly than does the live tissue. For example, in young colonies of Wreath coral, *Goniastrea aspera*, consisting of only a few polyps, a figure of 0.15 g $CaCO_3$ was obtained for a polyp, whereas in old colonies composed of some 500 polyps, it was 0.80 g $CaCO_3$ (Kawaguti, 1941). Since *Goniastrea* colonies are more or less spherical, their volume increases according to the cube, but the area according to the square of the radius. In colonies non-spherical in shape, the ratios shift in favour of the live tissue, and disc-shaped colonies grow relatively more quickly than compact ones, branching more quickly than disc-shaped.

Not only the shape of the colonies leads to differences in the rate of growth, but also the texture of the skeleton. According to their structure, a distinction can be made between the Perforata and the Imperforata. In the Perforata, the limestone network linking the individual corallites of a colony is riddled with perforations. This "light-weight construction" leads to a relatively high rate of growth with only minimum expenditure of material. For this reason, shrub or bush-shaped colonies have the ability to regenerate very rapidly after being damaged, which explains the predominance of Staghorn corals in particularly endangered shallow-water zones.

The age of the colonies also plays a part. Young colonies generally grow more quickly than old ones. Some alter shape during growth. Plicate coral, *Manicina areolata,* grows rapidly until its fourth year, then progressively more slowly. At four years, the colony measures 6 cm, at 8 years 8–10 cm. Certain ecological factors also affect the growth of coral colonies, so that it is much greater when conditions are optimal than when they are just tolerable. Wrecked ships often provide good points of attachment. On one ship sunk in 1792 off the coast of America, that lay at a depth of 7.5 m, an Elkhorn coral, *Acropora palmata,* is said to have reached the amazing height of 5 m after 65 years. If these rather old data are reliable, the rate of growth must have been 7.7 cm a year. This corresponds approximately with the findings of Jones who recorded an annual increase in length in Indo-Pacific species of *Acropora* of not quite 10 cm. An annual growth of as much as 26 cm has been measured for the Staghorn coral *Acropora cervicornis.* A ship sunk in 1933 was still free of coral growth after 2 years, but after 27 years, Bulbous Pore coral, *Porites astreoides,* had grown to a diameter of 12.5 cm, and Caribbean Brain coral, *Diploria strigosa,* almost 9 cm.

From measurements of this kind, an attempt was made to deduce the speed of growth of a reef. With an average vertical growth of 6 mm per year for the Small Button coral, *Montastrea annularis,* a coral reef will reach a height of 45 m after 7,500 years (Vaughan, 1916). Rapidly-growing Elkhorn coral could build such a reef after only about 600 to 1,000 years. Although such calculations are purely speculative, it is nevertheless illuminating to inquire into the share taken in the building of a reef by the individual species of coral with their different rates of growth, different degrees of rarity, differing sizes and weights. Kühlmann calculated that in the Cuban coral reefs, the contribution made by Elkhorn corals (*Acropora palmata*) to reef-building was some 2.6 times greater than that of the Small Button coral (*Montastrea annularis),* 6.0 times that of Staghorn coral (*Acropora prolifera*) and 12.4 times that of Brain coral (*Diploria strigosa),* which, together with a few other species, are among the major reef builders in this region (Ill. 90).

Growth is not uniform but depends upon a phenomenon characteristic of all organisms, namely biorhythmic intervals. In hermatypic corals, they are determined principally by processes of assimilation and dissimilation in the zooxanthellae that are dependent upon the position of the sun, producing much more intensive calcium carbonate crystallization during the day (see p. 34). Different metabolic effects resulting in this way are reflected in extremely slight daily increase rates that are visible by radiography. From similar features in fossil corals, it has been deduced that 350 to 400 million years ago, the earth revolved round its axis more rapidly, namely once in 21 hours, the year had 420 days and every 50,000 years, the day increased by a second. In this way, corals confirmed the astronomical clock of the geophysicist.

Currents, sand and sun

No living organism is able to exist independently of its environment. Sessile Scleractinia are particularly subject to it because they are unable to evade its rigours. The intimate connection between hermatypic corals and physio-chemical factors drew the attention of men of science long ago. More than 200 years ago, Georg Forster wrote: "As far as I have been able to observe, those coral reefs that exist on the side towards which the wind blows are, for the most part, the tallest and the most luxuriant." The demand for fairly turbulent water which some rapidly-growing corals make on their environment is not a universal one. On his circumnavigation of the earth, Darwin observed the differences exhibited by species of coral in tranquil lagoons and those on the margin of the outer reef, exposed to the action of the surf. A sure criterion of the acceptibility of the living conditions is the prevalence of reef-building. It takes place only in clean clear water with a salinity of 27–41 per cent at temperatures that do not fall below 18°C, with an intensity of light such as is found at a depth of about 40 m, where there is a constant exchange of water and a rocky bed. These features represent the optimal conditions. What is their effect in detail?

The salinity of the sea is predominantly 34–36 per cent, and can be considered as a very stable factor. Evaporation or the influx of fresh water cause some degree of local fluctuation above or below the norm. In the Red Sea, where virtually no free exchange of water with the ocean exists, extensive evaporation produces a salt content of 38–41 per cent. The increased salinity does not impede the growth of corals that develop in profusion here. A similar phenomenon has been observed in the enclosed atoll lagoons.

A slight reduction in salinity sometimes follows heavy tropical rainstorms, but it is restricted to the upper layers of the sea, because water with a lower salt content is lighter than that with a high one. So rain water remains on the upper surface of the ocean, and even corals growing no more than 20–30 cm lower down are scarcely affected by it. Common shallow-water species show a

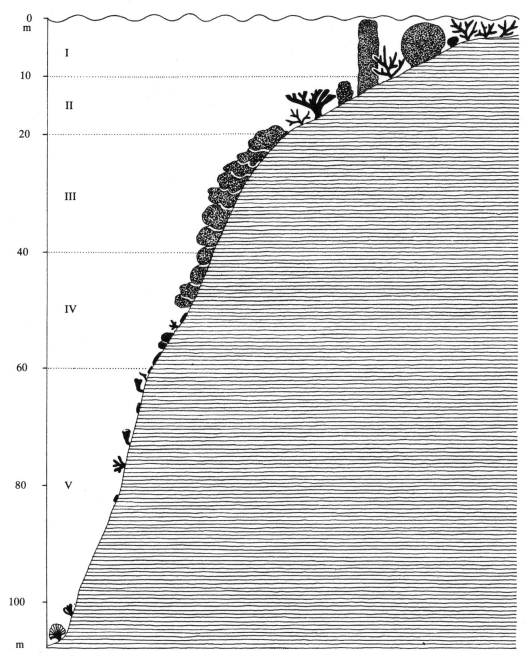

shower of rain falls during their brief exposure to the air, they will die as a result of saline deficiency.

The temperature of the oceans in the tropics fluctuates very little. In large water masses, marked variations are found only where cold currents come in from the Antarctic; this happens in the Pacific Ocean off the west coast of South America, in the Indian Ocean off Western Australia and in the Atlantic off Namibia and Angola. Although these regions of the ocean extend well into the tropics, they are too cold to permit an intensive growth of hermatypic corals. No reefs exist there. Off the western coast of South America, the effect of the cold Humboldt Current is so strong that it is only in the Gulf of Panama that the first sparse coral reefs are found, that is, beyond the equator. In these reefs, the growth rate of Claviform coral, *Pocillopora damicornis,* is reduced by about 20 per cent.

On the other hand, warm currents can carry the activity of reef-building far beyond the tropics. The Bermuda and Hawaiian Islands provide examples of this. In the reef areas, temperatures vary within the limits acceptable to the corals, namely between 18°C and 30°C. But individual corals demand different temperatures, and even within a single species, the tolerance range can be altered if the colonies had their origins in different ambient temperatures. For example, in the case of Claviform corals *(Pocillopora damicornis)* and Micropore corals *(Montipora verrucosa)* from the warm waters of the Eniwetok atoll, the lethal temperature was found to be as high as 35–38°C, while for members of the same species on the reefs off Hawaii, it was 31–33°C—undoubtedly a response to the colder waters round Hawaii. Below 2–3 m, the water temperature is very constant. For many species, a temperature of between 25 and 31°C can be considered optimal, since this is the most favourable for the precipitation of calcium.

10   Stony coral growth in the various photic zones. I "Red zone" with predominantly bushy corals. II "Yellow zone" with massive and bushy corals. III "Green zone" with leaf-shaped corals. IV "Blue zone" with leaf-shaped and loose, encrusting coral growth. V "Dark zone" with scanty growth of aposymbiotic corals (Original)

particular degree of resistance to salinity fluctuations, and they are precisely the species most subject to it. In tidal zones, the corals of the reef ridge can even be freely exposed to the sun at low tide. Abundant secretion of mucus and the retention of water in the hairlike channels of the skeletons prevent dehydration, and the corals are able to survive for an hour or two, until the tide rises to cover them again (Ill. 26). But if a violent

37

One of the most important factors in the life of reef-building corals is light. As it penetrates water, it undergoes a triple modification. Part of it is reflected as it strikes the surface. Within the water, it is dispersed more and more with increasing depth, becoming so diffuse that soon, objects cast no shadow. Thirdly, the water acts as an optical filter. Even distilled water absorbs selectively in descending order first the longwave, then the middle and finally the shortwave spectral region. In sea-water, this process is intensified by the quantities of matter in solution. It is for this reason that the brilliant red colouring of many organisms cannot be perceived, even at a depth of no more than 10 m. By 20 m, the yellows, by 30 m the greens and by 50 m even the light to mid-blues have been filtered out. Visual conditions can also be altered by the degree of turbidity of the water and its own inherent colour. The underwater photographer regains the full colour spectrum by using an electronic flash.

Vertical zoning exists in so far as the colonies of corals in shallow water differ from those lower down (Fig. 10). It was natural to assume at first that the selective absorption of the spectrum with increasing depth was responsible for this. However, the vital zooxanthellae, like higher algae and vascular plants, produce forms for conditions of both light and shadow, as has recently been demonstrated in the Button coral, *Montastrea*

annularis. The "shade" forms possess more of the pigments, chlorophyll and carotenids necessary for photosynthesis than do the "light" forms, and in this way compensate for the lack of light at greater depths, by making better use of the available spectral components. In this way, maximum photosynthesis is maintained. This light-adaptive mechanism enables the majority of reef-building species of coral to live in depths of 0 m to 40 m and more.

Between 30 per cent and 3 per cent of the sunlight that strikes the surface of the water is the optimum for the coral's symbiotic algae. Metabolism can be reduced by up to 50 per cent by cloud. Accordingly, a luxuriant growth of corals with many species, indicating good environmental conditions, is found at a depth of between 3 and 30 m. Depending upon the extent of turbidity and the water's natural colour, it may lie higher or lower in the water. In the reefs of Bimini in the Bahamas, the most luxuriant growth occurs at 12 m. In the clear waters of the Gulf of Aqaba, the number of species increases progressively to a depth of 30 m, and off Jamaica, it only starts to decrease below a depth of 40 m. Parasymbiotic, usually solitary corals, often very beautiful in form, live in depths of up to 5,800 m in the dark, cold waters of the ocean. These facts indicate that because of algal symbiosis, hermatypic corals are all dependent upon light, but that light, since it is present in an

adequate quality and quantity to a depth of 40 m, is also a very stable factor. It does not limit individual species to particular depths, but limits all hermatypic Scleractinia to the uppermost layer of that zone of the ocean that is accessible to light, the photic zone. Other factors are responsible for the zoning into particular coral communities (see p. 43).

Every day, tons of soil, loose clay and sand are carried into the sea by rivers. These sediments settle upon the corals and cause them great damage. They also cover the rocky bed and occupy potential settlement areas. Silt, then, is a deadly enemy of corals. For this reason, no coral reefs have developed off the mouths of major rivers, the Mississippi, Amazon, Congo or Ganges. Nor are they found near the mouth of a smaller river, and where tracts of reefs stretch along a coast, there are gaps off the river mouths.

But stony corals are not entirely helpless in the face of sedimentation. They counter it by the shape and orientation of the skeleton, and by active cleaning mechanisms. Steeply-domed skeletons, and even more so branching ones, offer little opportunity for sediment to settle, and are easily washed clear by movement of the water. This, of course, introduces an element of selection. In the Gulf of Aqaba, three main species of coral grow in the zone of maximum sedimentation, *Stylophora pistillata* (Styloid coral), *Acropora hemprichi* and *A. varia-*

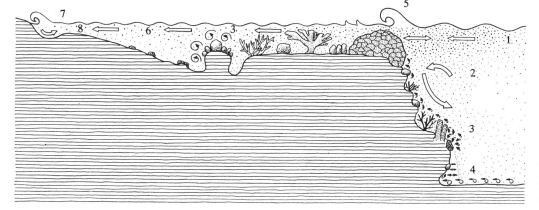

11 *Water movement and sedimentation in a bank reef. 1 wind waves and wind drift, 2 current towards slope, 3 turbulence, 4 barrier-effect dam-up of water, 5 surf, 6 surf overflow, 7 inflow to shore, 8 backing-up of water at shore (Original)*

bilis (Staghorn corals). They are bush-like growths and can resist silt deposit without difficulty. Fine structures are also significant. In certain species of Mushroom coral, Schuhmacher showed that the narrower, the more groined, the sharper and the more serrated the margins of the septa, the lower is the resistance to sedimentation. The weight of silt presses the sensitive cell tissue of the polyps against the sharp edges of the septa and it is cut. But if the edges are rounded, no damage is caused.

Stony corals have evolved various strategies by means of which they actively clean their surface.

1. Corals that lie freely on a sandy ground, such as Plicate corals *(Manicina areolata)* and some Mushroom corals *(Fungia actiniformis, Diaseris distorta)* can dig their way out, even when totally covered in silt, by repeatedly inflating the body. Certain smaller species, such as *Cycloseris,* turn somersaults and free themselves of sediment in this way. All these corals are able to live in habitats with a high rate of sedimentation.

2. Species with large polyps pump themselves full of water, swell up like bladders and let the silt slide off. Many Plicate corals *(Symphyllia)* as well as Button corals *(Montastrea annularis* and *M. cavernosa),* Atlantic Thick-stemmed Umbellate coral *(Mussa angulosa)* and Tuft coral *(Dichocoenia stokesi)* protect themselves in this way from being smothered in sand.

3. Small-polyped Pore corals (Porites) remove particles of dirt by using their tentacles to pass them from polyp to polyp, until finally they fall clear of the colony.

4. The beating action of cilia, already mentioned above (see p. 23), is usually directed away from the mouth. In this way, fine sediment is conveyed from the oral disc to the edge of the colony. This method of transporting material has been observed in many species such as Staghorn and Fluted corals, Brain corals of the genus *Colpophyllia,* Plicate, Star and Meandrine corals, and seems fairly widespread. In those corals with long tentacles, the ciliary current is directed permanently outwards. If a fragment of food is to be brought to the mouth, the tentacles are simply curved inwards without altering the direction of pulsation, and the material is carried in the opposite direction.

5. Many corals, including Mushroom and Plicate corals and the Funnel coral, *Siderastrea radians,* envelop dirt and silt particles in mucous secretions and, by making use of the cilia, carry them away on a track of mucus.

The extent to which the Scleractinia are able to free themselves from dirt and sand particles is varied, and determines their susceptibility to processes of sedimentation (Ill. 28). But no coral is able to withstand the effects of extremely heavy silting associated with volcanic eruptions, flooding or human activities.

Water movement in different forms and degrees of intensity has considerable and highly differentiated effects on stony corals. In 1849, Alexander von Humboldt wrote: "According to the experiences of Ehrenberg and Chamisso in the Red Sea and among the many atolls of the Marshall Islands east of the Caroline Islands, and according to the observations of Captain Bird Allen and Moresby in the West Indies and the Maldive Islands, living Madrepores, Millepores, Astreas and Meandrines can tolerate the roughest action of the waves, indeed, they even seem to prefer exposure to the breakers." And it is true that although they cannot withstand the excessive violence of hurricane-whipped seas, stony corals do in fact prefer turbulent water. In the literature, the reasons given are the constant renewal of the supply of water rich in plankton and oxygen, and the absence of sediment. Undoubtedly of equal importance is the removal of excreted products of metabolism. In shallow lagoons that heat up during the day, and in which extensive coral beds are found, waves breaking across the ridge of the reef or currents making their way through channels in the reef into the lagoon also have the effect of regulating the water temperature and ensuring a constant renewal of water (Fig. 11). The consequence is that with increasing intensity of water replacement, there is denser growth of prolific species.

But here too, extreme events inevitably lead to damage. Somewhere thousands of metres below the surface of the sea, tremors shake the earth. The underwater convulsions that are often accompanied by volcanic eruptions are transferred to the water masses and burst out of the ocean as tidal waves of unimaginable power. These tsunamis sometimes engulf whole villages and islands.

With their deep, thrusting waves they strip the reef slopes bare to a depth of 20 m or more, throw corals up onto the shore, piling them into banks of debris several metres high (Ill. 24, 25). Yet the devastated rocky beds are quickly recolonized by corals, and only two decades later, are again covered with a luxuriant growth.

In experiments, planula larvae released from a parent polyp have been found to be fairly selective in choosing suitable ground on which to settle. From the very start, this substratum is an important factor in the life of the coral. The great majority of species settle on rocky ground. Since this substrate is frequently occupied by algae, sea urchins such as *Heterocentrotus mammillatus* and *Diadema setosum* that browse in large numbers on algae, can act as path-clearers for colonization by corals. Differences in form and weight of the coral skeletons mean that not every available rocky bed is suitable for all species. Massive, heavy colonies grow on a horizontal substrate. On flat or slightly inclined ground, bush-shaped corals that grow in hedge-like colonies are also less liable to be damaged. Encrusting or disc-shaped colonies can cling to steeply sloping or vertical rock faces. A whole series of species colonize both steep and level ground, but in either case produce a different form of growth. A few specialized corals live on sandy ground. Since in a coral reef, level stretches of rock alternate with slopes, and steep faces with caves, and even quite large structures are subdivided by shelves, edges, projections, gaps and hollows, the substratum is principally responsible for the great diversity in the coral colonies, which increases with the increasing fragmentation of the reef surface.

Plastic rigidity

Salinity, temperature and light are all reasonably stable features in tropical reefs. Unstable features are substratum, hydrodynamics and sedimentation; substratum, because the surface structure of a reef alters from one metre to the next, hydrodynamics because waves and ground swell are variable and, depending upon the exposure of the reef or part of the reef to the wind and the open sea, produce effects of varying intensity that diminish with increasing depth, and sedimentation, because it is determined by the quantity, nature and portability of the silt, causing windward and leeward sides to develop in reef areas. Sands are moved from windward to leeward side. Outer slopes of a reef are therefore free of fine sand, a condition greatly favouring the growth of corals. The close interaction of the three unstable environmental factors of substratum, hydrodynamics and sedimentation can be described as the mechanical complex of factors (Fig. 13). It brings about the thorough mixing of the water, so that temperatures, nutrients in solution and particles of food are evenly distributed and noxious products of metabolism removed.

Moreover, this complex of factors plays a decisive part in determining the species-characteristic forms of growth and controls the varied composition of the coral communities.

Ecological growth forms are the different but characteristic forms that a colony of stony corals can exhibit. Their development proceeds within a fixed pattern of genetic variability, but is set in motion and controlled by induction factors. Ecological growth forms or ecomorphs are known in scientific literature as "forma", which, abbreviated, becomes "f". Four growth forms are described for the Caribbean Elkhorn coral: *Acropora palmata* f. *palmata* has sweeping, spade-like forms. Because of its wide distribution and frequent occurrence, it can be considered as the "basic form". It is common in moderately to highly turbulent water, predominantly at depths of 1 to 5 m. *Acropora palmata* f. *erecta* shows upright, almost vertical growth throughout, with scarcely any branching. It grows in deeper, more peaceful waters. *Acropora palmata* f. *retroflexa* has almost horizontal branches, all tending in a particular direction. They follow the prevailing movement of the breakers. *Acropora palmata* f. *crustosa* covers the substrate with a relatively thin crust. It grows close to the surface of the water where there is heavy surf action, or in other conditions bordering on the minimum survival level, and it can be considered as the stunted form of the species.

Whereas the ecomorphs of Elkhorn coral are clearly linked to water movement, Fluted coral, *Agaricia agaricites*, shows different forms depending upon the substrate. On horizontal ground, it forms massive, hemispherical colonies. On edges of rocks, where it often grows densely at depths of 1 to 10 m, the colonies are foliaceous, interlocked with one another and covered on both sides with rows of polyps. On steep rock walls, it grows in discs of up to 1 m in

diameter, without polyps on the side facing the wall. It occurs only in calm water, and is common at depths of 20 to 60 m. In quiet lagoons, it may be found on a steeply sloping substrate at a depth of no more than 4 m (Fig. 12).

On the other hand, the ecological growth forms of Button coral *Montastrea annularis,* that grow off Jamaica, forming semispherical colonies, above a depth of 20 m, are dependent on the factor of light. Since a sphere combines the greatest possible volume with the smallest possible surface area, semispherical growth in shallow water can be interpreted as an indication of optimal limestone production, while the more flattened growth at greater depths can be attributed to the reduced intensity of light there; much less calcium carbonate is required to produce a flat, spreading colony than a semispherical one.

If the influence exerted by an ecological factor becomes very great, the coral colonies react by producing skeletal deformities. Striking phenomena of this kind are the "micro-atolls" frequently found in shallow water, sometimes with a diameter of several metres, colonies that are dead in the centre and that continue to grow only at the periphery. The lifeless centre is eroded, the zone of growth bulges outwards. The living substance encircles the coral colony like a ring, giving it the appearance of a tiny atoll. This form of growth is seen mainly in massive colonies that have been affected by excessive quantities of sedimentation, or that have been left dry for too long at low water (Ill. 27).

These examples raise the question of how far particular shapes of colonies are especially well adapted to different environmental conditions. Let us first consider the three principal forms: branching, compact and flat colonies. Because of the multiformity of hermatypic corals, they can easily be subdivided further, for example, the branching forms into finely and coarsely branching,

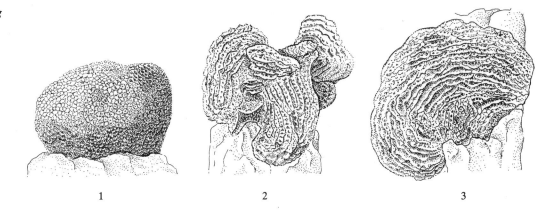

dendroid and columnar. The idea that optimal adaptation underlies this multiformity is confirmed by the development of ecomorphs. On the other hand, under the influence of like induction factors, different species exhibit similar colonial forms; in other words, specific forms of growth correspond especially well to particular environmental conditions.

A few examples will perhaps substantiate this. Colonies of Needle coral, *Seriatopora hystrix,* that flourish in calm water on sandy ground, have branches that are finer and slope more steeply upward than those growing on rocky ground. In this way they are better adapted to the heavier sedimentation associated with sandy ground, since silt falling through the water cannot easily settle on them. The many-branched Staghorn corals *(Acropora)* cause the water flowing past them to split into innumerable small eddies that swirl among their branches. In this way, all sections of the colony and all the polyps enjoy a vigorous supply of moving water. To a depth of 20 m, many species grow in semispherical or chunkily massive forms. Since water is continuously in motion round them and the prevailing conditions of light are favourable, their production of calcium carbonate is high. In contrast, steeply sloping faces are colonized by encrusting, slab-shaped or discoid stony corals, in which the centre of gravity is situated close to the substratum. As a result, they do not break

off. The capacity for ecomorph development can be considered as a perfected form of adaptation to various environmental conditions, and this plasticity which characterizes the rigid colonial forms, that may at first seem a contradiction in terms, is seen to be an ingenious adaptative strategy in response to widely varying environmental conditions (Ill. 29).

Doubly-sociable organisms

Reef-building corals are gregarious in two ways: they form colonies in which often thousands of individual polyps are united, and they come together in communities or associations in which different species grow close together within a small area. The formation of characteristic coral communities, depending upon specific environmental conditions, is an unmistakable feature, just as it is in plants. And just as there are many plants with a wide adaptive range, there also exist among stony corals relatively opportunist species that occur almost everywhere in a reef, as do Brain coral *Diploria strigosa,* Small Star coral *Favia fragum,* or Pore coral *Porites astreoides* in the Caribbean. But many species require special conditions in which to live, and occur only where their needs are satisfied. They are known as specialists.

Coral associations consist of three components: 1. characteristic species that make different demands on the environment, 2. characteristic forms that are composed of ecological growth forms of opportunist species, and 3. associates, opportunist species or specialists that have found their way into the habitat by chance, in which case the latter are not able to exist there permanently. Since the complex of mechanical factors (see p. 40) that most significantly determine the development of associations affects the various sections of a reef in a variety of ways, different coral associations are formed. Some examples from the reef region of the Caribbean will illustrate this.

The coral association of the surf zone grows on a horizontal or slightly sloping hard substrate that is accessible to a turbulent sea with moderate surf. Depending upon the violence of the moving water to which it is exposed, the depth at which it grows can vary between 0 m and 20 m. Waves, surf or ground swell make their way across an obstacle course consisting of a dense tangle of corals. Two characteristic species belong to this association. Fire coral, *Millepora complanata,* grows in places where the water movement is somewhat slighter or in the protection of the thick-stemmed Elkhorn coral, *Acropora palmata.* Sometimes they are accompanied by Brain coral *(Diploria strigosa),* Small Star coral *(Favia fragum)* and torose

Pore coral *(Porites astreoides)*. All these species are very resistant to coarse sand carried along in the water.

The coral association of the outer slopes enjoys very favourable conditions. The substrate is moderately sloping and richly varied in structure. The currents and eddies that stream across it lead to a constant exchange of water and prevent heavy sedimentation. Since the associations are usually found at a depth of 2 to 20 m, optimal light is available to them. A large number of species settle here, and it is notable that there are really no characteristic species or forms. Button corals *(Montastrea annularis* and *M. cavernosa)* grow beside Tuft corals *(Dichocoenia stokesi)*, Brain corals *(Diploria strigosa* and *D. labyrinthiformis)* beside Candelabra corals *(Dendrogyra cylindrus)*, Plicate corals *Mycetophyllia lamarckiana* and *Isophyllia multiflora)*, Cactus, Bouquet, Funnel, Pore corals and others all grow together in rich confusion. Species of the surf zone also find acceptable conditions here in the upper parts. The highly varied nature of the ground, with its ledges, niches, terraces, caves and steep faces causes a certain degree of differentiation, because many species prefer particular angles of inclination.

The coral association of moderate water movement can occur twice in one reef, once at the edge of the reef on the lagoon side, where the action of the surf has lost much of its force, and again in the deeper, calmer zones of the outer slopes, on a horizontal or slightly sloping rock bed. In both habitats, Staghorn corals *(Acropora prolifera* and *A. cervicornis)* build extensive, dense hedges. Beside them, finger-shaped Pore coral *(Porites porites)*, Brain coral *(Diploria strigosa)*, Small Star coral *(Favia fragum)*, Button coral *(Montastrea annularis)* and the fine Fluted coral *(Agaricia agaricites)* are frequently found. This coral association has features in common with that of the outer slopes, since the

quality of the water movement scarcely differs, being merely somewhat milder here.

The coral association of steep rock faces develops under very different conditions, usually below a depth of 20 m. It is difficult for the coral colonies to maintain a hold on steep faces, so they grow close to the rock (Ill. 40). The picture this coral association presents is therefore a very uniform one. Hydrodynamic processes, although reduced, are not obstructed entirely, and slight, scarcely perceptible currents ensure a continuous movement of water. In this situation, powerfully moving water would simply accumulate in front of the flat-spreading coral colonies with their vertical tendency of growth, producing a barrier effect that would prevent the exchange of water and the introduction of fine nutrient organisms. Characteristic species are Fluted corals *(Agaricia fragilis, A. lamarcki, A. agaricites* f. *purpurea)*, Wave coral *(Leptoseris cucullata)*, Tessellated coral *(Madracis formosa)*, flat Plicate coral *(Mycetophyllia reesi)* and Tuft coral *(Dichocoenia stellaris)*. They are accompanied by Encrusting corals *(Stephanocoenia michelini)* minute solitary Stump corals *(Astrangia solitaria* and *Phyllangia americana)* and on small projections, Meandrine corals *(Meandrina meandrites)* and Large Star corals *(Scolymia)*.

An interesting phenomenon is the coral association of seagrass or eel-grass meadows. Here, in the calm water of the lagoons, where suspended matter is abundant, the corals have adapted to the sandy ground that has been consolidated by eel-grass and is protected against strong drift currents. The corals in question are Tube coral *(Cladocora arbuscula)*, the small branching Pore coral *(Porites divaricata)*, Plicate coral *(Manicina areolata)* and Funnel coral *(Siderastrea radians)*. All species living here have the ability to free themselves actively from sedimentation by means of

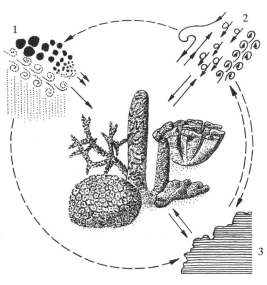

13 The effect of a complex of mechanical factors on corals. 1 sedimentation, 2 hydrodynamic conditions, 3 substratum (After Kühlmann, 1978)

the cilia and tentacles (see p. 39). In addition, there are various features in the form of their exclusively small colonies that make it possible for them to live on sandy ground. The limestone skeleton of the branching Tube coral is so light that it sinks scarcely at all into the soft ground. The small branching Pore coral also enjoys this advantage. Its tiny branches grow upwards. But at a certain time, the colony turns over, and new branches grow upwards. The base becomes progressively larger and more stable. At the same time, its fragility increases, and one day the colony falls apart. The separate parts develop into new colonies—a form of vegetative propagation. The round to oblong-ovate Plicate corals that also grow occasionally on hard ground, only just peep out of sandy ground. The underpart of the colony is conical and provides anchorage in the ground. They lie like discs on the coarse sand. A particular mode of adaptation has been evolved by the Funnel coral. It usually grows in a more or less clearly domed shape on rock. In the lagoon, it either makes use of other solid objects such as snail shells or any available pebbles as a base on which to settle or forms spherical colonies. These are rolled about by the action of the waves. Thus different parts alternately come in contact with the ground, and all the polyps are able to enjoy the open water in turn.

Convergent forms also occur in the Indo-Pacific. The finely articulated, branching Needle coral also sinks down into sand, *Trachyphyllia* inserts its conical-shaped end into sandy ground, and *Porites* and *Cyphastrea* also form spherical colonies on sand. So adaptive forms of various corals capable of living on sandy ground have universal distribution. But the Indo-Pacific Hat corals (*Halomitra* and *Parahalomitra*) have no counterpart on the Atlantic shelf. They build large, deep, bell-shaped colonies. The skeleton is covered both internally and externally by live tissue. Even when the coral colony is lying with the rim on the ground, there is no danger of the water underneath the bell stagnating, since the skeleton is fine, porous and pervious to light, providing no impediment to gaseous exchange.

In similar environmental conditions, associations analagous to the Caribbean coral associations occur in the Indo-Pacific. But a number of peculiarities make the overall picture less clear. For example, the Red Sea exhibits altered hydrographic conditions because of its isolation. Many reefs in the Indian Ocean and the Pacific are subject to strong tidal movements, others to powerful ocean swell. In the Indian and Pacific Oceans there are many coral islands at a very great distance from the mainland and well beyond its sphere of influence. Geographical barriers probably also lead to special characteristics. Because of the vast size and diversity of this region, we do not yet know enough about it to venture exact statements on the coral associations that exist there.

But one feature seems to be characteristic of the coral reefs in all oceans and

seas, that is, the creation of outwardly similar zones of growth.
— The zone of bush-like growth is frequently seen in shallow water on a flat to slightly sloping substrate. Here we find predominantly branching stony corals, such as Staghorn, Claviform, Styloid corals and branching species of Pore corals (Ill. 31).
— The zone of block-like growth can develop on a terrace at a depth of about 5–10 m, or on the outer slope of a reef. Here, huge colonies of Pore corals, such as *Porites lutea,* tower up almost to the surface of the sea in the form of blocks, the so-called "bommies", or tower-like coral parapets 5–10 m high. Exceptionally large colonies of Star, Button, Funnel or Brain corals often make up the sides of the reef.
— The zone of bulbous growth is found on slopes at a depth of 2–25 m. It is a zone of confusedly intermingled accretion, in which a multiplicity of species occurs in the form of massive or hemispherical colonies. This zone often overlaps with the zone of bush-like growth (Ill. 45).
— The encrusting zone exists principally on steep slopes at depths of 0–10 m. Representatives of Pore, Micropore and Tessellated corals form inconspicuous encrustations, Honeycomb and Star corals striking ones.
— The zone of discoid growth usually occurs at a depth of 20–60 m. It is made up of three forms of growth that differ according to the angle of inclination of the substrate, but which are in each case flat. On steep rock faces, the vertically growing disc-shaped colonies of Fluted, Undulating, Plate and Gramophone-record corals predominate. On more or less steeply sloping faces, Star, Pore or Button corals have been found arranged one over the other like tiles on a roof. Foil and Micropore corals

form rigid spirally-twisting c
(Ill. 47).
— Finally, mention should be made oi the zone of palisade growth that belongs to a depth of 10–20 m, and seems to be a local phenomenon, occurring only in the Indo-Pacific at the foot of reef slopes. Here, pillar-like colonies of Scalpel coral, often more than one metre in height, stand packed close together.

The individual environmental factors can be classified according to their differing functions. 1. Ecologically stable features such as salinity, temperature, oxygen content and food determine the limits of geographical distribution of hermatypic species of coral, and can therefore be called ecogeographical factors. 2. The factor of light, which sets the limits of depth and the extent of the reef-building processes for all reef-building corals, also appears to influence the growth forms through the processes of limestone formation. It varies in direct proportion to the depth, and can be called the ecobathymetric factor. 3. The factors of hydrodynamics, substrate and sedimentation vary in quantity and quality in even a very small space. Since corals react very sensitively to them, they are the factors mainly responsible for the composition of the coral associations, making changes that alter the overall picture; they are ecotopographical factors. Moreover, because they continually influence and alter one another and therefore always have a combined effect on corals and other sessile organisms, they can be linked as a complex of eco-mechanical factors (Fig. 13).

However varied the conditions of life, however diverse the mutual interactions, and however complex the links of relationship, corals as sessile organisms have had to face up to all these factors, and they have adapted to them so admirably that they are able to exploit them to their own advantage.

reef-builders and their work in the reef

In addition to corals, many other limestone-secreting organisms contribute to the building of the reef. Who is not familiar with the armoured crustaceans, the mussels that shimmer inside like silk, the snails in their elaborately turned shells? Very often, the limestone skeleton weighs more than the soft parts of these creatures. When they die, the hard substances are usually crushed and incorporated into the rock of the reef. But a much higher proportion of the building material for the coral reefs is provided by sessile organisms. They live on every rock face, on every ledge, in every niche. Because they demand a similar environment, they live among stony corals. Even the dead underparts of the coral colonies are colonized by them. As a result, every centimetre of substratum in a reef is covered with organic growth, and pieces that are broken from it are aptly described by aquarists as "living stones".

Calcareous algae

Lime-depositing algae or coralline algae are mainly restricted to tropical oceans. This is an advantage since the solubility of calcium carbonate decreases with increasing temperatures. The algae deposit lime in the form of aragonite or calcite between the cell walls or outside the thallus. They are clumpy, crusty, some species branching. The clumpy or crusty ones are so inconspicuous that they can easily be mistaken for stones

(Ill. 37). *Halimeda,* calcareous Green Algae, common in all reefs, have an articulated structure. The thallus consists of small, highly calcified green leaflets that are arranged side by side, each frond giving off side branches. Frequently they grow in such quantities on the dead base parts of stony corals that in places, their laminae are the main constituent of the reef sand.

Coralline algae do not represent a separate systematic group but occur as 590 species of red alga, 90 species of green and 2 species of brown alga. They are very dependent on light, water movement, substratum and the competition of other sessile organisms. Grazing marine organisms pose a particular threat to them. These factors cause considerable variation in the occurrence of species in the reef, and most species show preference for particular depths. Whereas red coralline algae live only on rock, a number of limestone-depositing green algae also grow on sandy ground. As in the case of stony corals, water movement causes the development of different growth forms within the species.

Rate of growth is lower than that of stony corals. A monthly increase in size of 0.9–2.3 mm has been observed in the encrusting *Neogoniolithon* (Adey and Vassar, 1975). Branching forms achieve a growth of 2–4 cm per year (Johannes, 1969). Growth is extremely slow at great depth, where the age of a disc of calcareous algae with a diameter of 20–30 cm, is estimated to be 500 to

800 years (Adey and Macintyre, 1973). Rapid growth is a feature of green algae that encrust their thalli with limestone. For example, *Halimeda* grows 4 cm in 36 days (Colinvaux et al. 1965). The alga *Penicillus* lives for only 30–60 days, but reaches a height of over 10 cm.

Calcareous algae have a two-fold significance for the coral reef. They bind and consolidate loose material, and they provide a quite significant proportion of the calcium carbonate that makes up a reef. In the reefs of Belize in Central America, this proportion is 20 to 25 per cent. On the edges of the Hawaiian reefs, the encrusting alga *Porolithon onkodes* that occupies 41 per cent of the surface, produces 0.5 g $CaCO_3$ per m^2 per day. Together with *Porolithon gardineri* which is equally common there, it consolidates the exposed reef margins so effectively that they are able to withstand the crushing force of the breakers. Indeed, heavy surf is an advantage to these algae since other organisms such as stony corals or gulf-weed *(Sargassum)* are unable to live there and are not therefore in competition.

Fire and Filigree corals

In reefs, it is not unusual to find growing upright from the rocks, rigid brown or yellowish structures that bear a close resemblance to true corals (Ill. 29). The hard limestone skeleton has minute holes or pores scattered over the surface. Seen against the light, it reveals an extremely delicate, almost transparent coating of cobweb-fine polyps. The diver who touches one experiences a violent burning sensation and understands at once why they were given the name of Fire coral *(Millepora)*. They belong to the class Hydrozoa, in which they hold a special place on account of their ability to build a limestone skeleton, a feature that has earned the name of Hydrocorals for the various representatives of this group. They also live in symbiotic association with zooxanthellae. The hair-like polyps, armed with batteries of extremely effective stinging cells, occur in two forms. The short, cylindrical feeding-polyps have four small polyp heads at the sides of the oral orifice. The large numbers of slender defensive-polyps form a circle round the feeding polyps. They are equipped with a varying number of stalked polyp heads. Both have very short tentacles (Fig. 16).

The polyps differ from those of the Anthozoa in that the gastrovascular cavity is smooth-walled and non-septate. A remarkable feature of most Hydrozoa is the alternation of generations or metagenesis. Sessile, colony-forming generations of polyps that reproduce vegetatively alternate with vagile (free-moving) generative medusae that live as solitary organisms. The Fire corals also produce medusae. They develop in special pouches in the tissue of the colony that is permeated by numerous canals, known as stolons, initially in the form of polyp-like creatures which later free themselves and move off as minute jellyfishes. They are transparent and almost invisible in water. Without tentacles or mouth, they function solely as carriers of reproductive products and ensure their distribution.

The class Hydrozoa also includes the Filigree corals that belong to the family Stylasteridae. They also deposit hard, limestone skeletons consisting of aragonite and are predominantly small,

branching and delicate, white, violet, yellow and red (Ill. 39). They also produce dimorphous polyps. A large number of defensive polyps, each consisting of a single tentacle equipped with many nettle cells, surround a feeding-polyp. So closely do they resemble stony corals that formerly they were assigned to the same systematic category. Filigree corals are heteroicous and produce no medusa generation. Sexual products are developed in special chambers that in many species are visible externally as nodules. They grow in dark places, in caves, in the many cavities in the basal part of stony corals or in quite deep waters. Inexperienced divers often mistake a red Filigree coral for Precious coral.

Frequently the density of aggregations of limestone-depositing Hydrozoa colonies in a fossil reef indicates that they were major participants in its construction. Although hydrocorals have not played such a significant part in the building of recent reefs, they are nevertheless not rare. In many of the reefs of the western Atlantic, milleporides are still today numbered among the most important sources of calcium carbonate.

Organ-pipe corals

Curiously constructed, intensely red Organ-pipe corals *(Tubipora)* are found quite commonly on beaches between East Africa and the Pacific Islands. Brought home by seafarers, they were already to be found in European Natural History Collections of the Middle Ages. Linnaeus, who described them more than 200 years ago, placed them in the stony coral genus *Madrepora,* into which he had gathered together a wide variety of corals, and assigned them to the large group of Lithophyta or Stone Plants. They consist of many separate bright-red limestone tubes arranged in rows and connected at intervals by horizontal platforms (Fig. 14). They resemble the pipes of an organ. Great quantities of skeletal spicules are produced in the supportive lamella of their tissue. These fuse

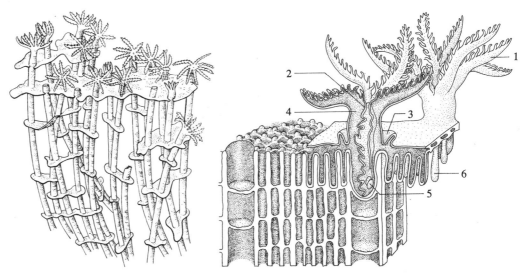

14 *The red skeletons of Organ-pipe corals,* Tubipora, *consist of separate parallel tubes of calcium carbonate linked together by transverse strata. (After Anderson, 1976)*

15 *Blue coral,* Heliopora coerulea, *has a robust limestone skeleton reinforced by longitudinal and lateral bracing. The living tissue extends saccular outgrowths into the upper chambers. 1 tentacles, 2 pharynx, 3 collar like fold of skin, 4 radial musculature, 5 ova, 6 pouch-like outgrowth of tissue (After Bouillon and Houvenaghel-Crevecoeur, 1970 and Grasshoff, 1981, modified)*

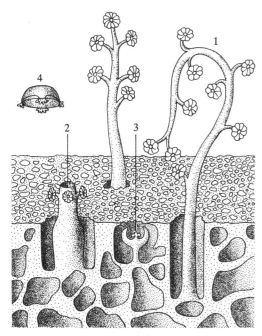

16 Fire corals (Millepora) *have a very firm, hard skeleton which, in addition to the channels that link the polyps, shows only minute chambers. Dactylozooids or defensive polyps (1) and gastrozooids or feeding polyps (2) differ in form. Growing inside the ampoules (3) are small medusae (4) that carry the sexual products through water and effect the distribution of the Fire corals. (Original)*

together to make the tubes that the polyps occupy, the upper portion of each being covered by the inner and outer skin of the animals. A complicated system of vessels branches through the connective plates, providing a means of communication between polyps. The creatures continually add to the height of the colonies and so avoid siltation. Organ-pipe corals grow to about the size of a man's head. Where they occur in large quantities, they provide a significant proportion of the limestone used in reef-building.

The flat parts of the reef are their habitat. The red skeletons are often concealed under a light layer of sand. The green polyps extend eight pinnatipartite tentacles from the tubes they occupy, and this feature provides the basis of their classification. Organ-pipe corals belong to the subclass Octocorallia, having tentacles in multiples of 8. Until quite recently, they were still classified within this group to the soft or leathery corals (Alcyonacea), but today they have been accorded the status of an independent order Stolonifera, because their elongate polyps stand quite separately and are not embedded within a common soft body.

Blue coral

Blue coral *(Heliopora coerulea)* has a light-blue calcareous skeleton (Fig. 15). Both this species and other representatives of the genus are found as fossils in Cretaceous and later formations along the Cuban coast, formerly the shores of the Tethys, through the European Alps as far as Japan and Indonesia. Today, *Heliopora coerulea* lives in the reefs of the Indo-Pacific. With some 100 million years of continued existence, it is a truly ancient "living fossil". Since its occurrence today is restricted to the hottest ocean regions between the Seychelles and Polynesia, it is decidedly designated as a heat-loving species. This is also demonstrated by its choice of habitat: the calm, shallow, sun-warmed water in the rearward part of the reef area. There, it usually forms clumps up to 50 cm high with finger-shaped or lobular processes, or if it extends to the low-water mark, micro-atolls over a metre wide. When its biomass is large, it makes a significant contribution to the reef.

The skeleton, constructed of aragonite crystals, exhibits a scattering of coarse pores on its surface, each occupied by a single polyp. It is covered by a layer 2 to 3 mm thick of living tissue, blue-green in colour. The minute polyps are often brown and their eight tentacles pinnate. They are linked together by a system of branching canals from which blind sacs project vertically downwards. Originally full of minute hollow spaces, the skeleton beneath the living layer calcifies completely. It becomes so hard that it can be cut and polished to make beads for precious jewelry.

Moss animals

Minute, net-like, whitish structures are often found on rocks, on mussel shells and seaweeds; they are the limestone skeletons of moss animalcules, the Bryozoa. The animals themselves are usually smaller than 1 mm in size, at most 2 mm. From openings in the skeleton, they project branching rings of tentacles that can be retracted at any time by paired muscles. By continuous whirling movements, they create a current in the water round them, which carries towards them food particles, ranging from very fine detritus to single-celled algae. The food passes through the esophagus, stomach and U-shaped intestine, and waste products are released through the anal aperture. A brain controls the creature's functions that are in part quite complicated. In addition to reproduction by budding and the formation of colonies, there is a considerable production of sexual products in the hermaphroditic animals from which larvae are produced.

Many of the separate individuals of a colony have been modified and specialized. Some of these zooids produce bristle-like processes. Others exhibit beaked or jawed structures that seize and hold organisms suitable for food. Already dead and decayed, particles of these organisms find their way into the tentacular currents and are directed into the animal's mouth. Other zooids have been modified as brood chambers in which polymorphous ciliated larvae develop. So in spite of their small size, bryozoans are surprisingly complicated animal colonies that have achieved a specialized division of labour by means of modification of the zooids.

The group has existed since the Cambrian period. The number of recent representatives is very high with some 4,000 species. Most of them live in the sea, predominantly at depths of 20 to 80 m, although some as deep as 8,000 m. The hard skeleton protects them from many enemies, indeed, so far, only sea urchins, chitons and nudibranchs as well as sea spiders of the genus *Pycnogonum* that feed by means of a proboscis, have been found to pose any real threat. To these can be added certain fishes equipped with powerful teeth that scarify the rocks.

The skeletons consist of calcite and aragonite crystals together with considerable quantities of magnesium and strontium. The colonies, that can be foliaceous, nodular or bush-like or in the form of encrustations (Fig. 17, 2), usually measure no more than a few centimetres, although sporadically they may grow to several metres. Although they occur universally, relatively few of them contribute to reef construction. But in past eras, they often played a dominant role.

Sessile, but elastic

In addition to rigid sessile organisms, elastic forms also occur in the reefs. Sponges are simply organized animals (Ill. 103–105). The basic sponge form shows a bulky cushion-like porous body. The pores lead in through a branching system of channels, by way of innumerable minute chambers to a broad central cavity, the spongocoel. From here, a large central aperture, the osculum, discharges outwards, rather in the manner of a chimney into which a large number of draughts enter from individual furnaces. The chambers are lined with collar cell flagella that ensure a constant flow of water. The majority of sponges form a skeleton of horny fibres in which unicellular algae (diatoms) and grains of sand are embedded. Many species support their skeleton exclusively by means of siliceous spicules that exhibit specific conditions of symmetry in their distribution and location (Fig. 19,1).

In tunnels and caves under the sea and in the deep-water regions of reefs, large populations of inconspicuous, stone-like sponges have been discovered by divers; in 1970, they were described by Hartman and Goreau as *Ceratoporella nicholsoni*. They deposit a massive limestone skeletal base upon which lies superimposed a thin layer of living sponge tissue, and in this, siliceous spicules are embedded (Fig. 20). They are the only organisms known to produce both calcium carbonate and silicate. Their unique structure has led to the establishment of the new subclass of coralsponges (Sclerospongiae). Anatomical and morphological characteristics clearly show a relationship with the fossil Stromatoporida (see p. 21). In all warm oceans, several more species of these coral sponges have already been discovered. Since they occur in massive quantities in deep-water zones, they often constitute a very important factor in the construction and consolidation of the basal plane of the coral reef. In the Upper Jurassic, some 140 to 150 million years ago, fossil sponge reefs reached considerable proportions in the Malm (White Jura) in the south of the German Federal Republic and in the Santonian stage at Saint-Cyr-sur-Mer in Provence.

Of much greater significance to reef building are two large groups of Octocorallia, the Soft corals (Alcyonacea) and the Gorgonians. Soft corals have an extremely thick rubbery supportive lamella. Embedded in the fleshy mass are large numbers of sclerites, calcareous spicules with rough, irregular surface and shape (Fig. 19, 2). In colonies that have fully unfolded and expanded, the supportive lamella is quite transparent and the large sclerites within it clearly visible (Ill. 61). Expansion is brought about hydraulically as specialized polyps, the siphonozoids, pump water into the colony. The colonies can resemble lobed, finger-like masses, succulents or branching plants, lumpy encrustations or giant mushrooms, and not infrequently reach a size of more than

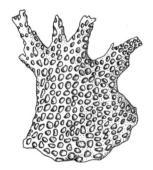
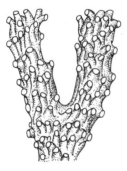

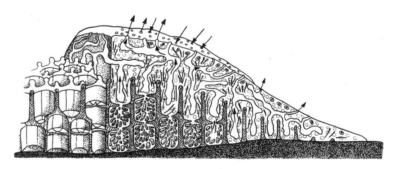

17 *Coral-convergent skeleton formation in reef organisms. 1 the sessile red foraminifer* Homometra rubra, *2 the bryozoon* Hornera sp., *3 the tubes of the polychaetes* Filograna implexa *that live in association (After Dana, 1875 and Schuhmacher, 1976)*

18 *In longitudinal section, the coral sponge,* Merlia normani, *shows two layers, the rigid limestone skeleton that has been deposited underneath, and the elastic upper layer that has many incurrent pores and is supported by siliceous spicules. The arrows indicate the direction of flow of the water currents. (After Kirkpatrick, 1911)*

one metre. They are covered with large numbers of polyps. After they have died, the sclerites are incorporated into the mass of the reef.

Gorgonians (Gorgonacea) develop a very firm flexible skeleton of a horn-like material. It is surrounded by layers of live tissue. Again, needle-like skeletal elements are secreted in the supportive lamella and laid down in a regular arrangement in the outer cortical layer. The high-tensile, axial skeleton allows the colony to grow in shrub-like or flat, net-like forms that can reach a height of 2 m. Particularly tough species grow among heavy surf. Their branches, that are spread in a single plane, are always extended at right angles to the prevailing movement of the water to obtain the best possible benefit from the inflow of plankton. The large numbers of polyps on every branch give them the appearance of being covered in flowers.

It is noticable that Soft corals are more common in the Indo-Pacific reefs, Gorgonians in the Caribbean, and that each in its own region is equally involved in the process of reef construction. In some of the Caribbean reefs, Gorgonians contribute more to the reef than do the Scleractinians. For example, in the reef area of Tortugas, south of Florida, 4 tons of skeletal spicules per hectare are released annually by dead Gorgonians lying at a depth of 2 m. Overall, the spicule mass from Gorgonians living in one hectare, at depths of up to 6 m, is estimated at 22 tons.

Tubes, shells and protective armour

Many living creatures find protection within tube-like coverings, shells and other natural housings. These refuges are usually constructed of limestone, and their inhabitants are numbered among the seemingly endless ranks of the coral-reef builders. Many marine ringed worms (Polychaeta), such as serpulids, build homes in the form of tubes made of limestone. It is secreted from a collar-like gland situated at the front end of the animal. The tubes are thick-walled and attached to a substratum; they can be spirated or irregularly curved, round or angular in cross-section, they may lie horizontal or stand erect (Fig. 17, 3). The worms never leave the tubes. They extend circles of feather-like tentacles with which they catch particles of food. If disturbed, they close the aperture by means of the operculum, a stalked lid or plate. They are found living in all coral reefs (Ill. 60, 95, 96).

In the calm water of the atoll lagoons, there are great quantities of bivalves (clams, scallops, oysters and mussels). Ark shells *(Arca)* and Thorny Oysters *(Spondylus)* are common. In some places, close-packed populations of Giant Clams of the genus *Tridacna* occupy the upper parts of a reef (Ill. 57, 58). This group includes the world's largest species of bivalve *Tridacna gigas*. Its extraordinarily thick valves can measure 1.35 m and weigh as much as 500 kg. It has also been called the "Killer Clam", since it is said to trap divers who inadvertently come between its valves, holding them fast until they drown. Certainly it will clap its valves together in response to stimulation, but an accident of this kind is very unlikely to befall an experienced diver who is conversant with the sea and its fauna. When a clam has died, its shell is incorporated into the substance of the coral reef.

Sea snails living in coral reefs usually lead a secretive life, searching out holes in the rock, hiding under stones, burrowing into sand or lying concealed under a covering of algae. But they are quite common, and anyone familiar with their habits will be able to find them. Many of them such as Helmet shells (Cassidae), Fighting conches (Strombidae), Top shells (Trochidae), Rock shells (Muricidae), Cowries (Cypraeidae)

49 14

14 *Micropore corals produce colonies in a wide variety of forms. Here,* Montipora verrilli *shows spirally twisted discs. (Tuamotu Archipelago, Pacific)*

50/51 15 16 | 19
        17 18 |

15 *The very large, fleshy polyps of Star corals* (Favia) *produce mainly compact colonies. (Red Sea)*

16 *Honeycomb coral* (Gardineroseris planulata) *often occurs in this compact form, but on steeply sloping walls it is encrusting or foliose. (Society Islands, Pacific)*

17 *The surface of Caribbean Brain coral* (Diploria strigosa) *clearly shows the fusion of individual polyps into long, continuous, meandering rows with convoluted depressions between. (Virgin Islands, Western Atlantic)*

18 *In Rough Star coral* (Echinopora gemmacea) *with a compact growth form, the raised polyp cups lie close together. (Red Sea)*

19 *The flattish discs of* Pachyseris speciosa *are characterized by deep grooves in which the polyp cups are embedded. (Society Islands, Pacific)*

52 20 21 |
        22 23 |

20 *During the day, Bladder coral* (Plerogyra sinuosa) *retracts the tentacles and inflates bladder-like organs. These contain zooxanthellae, which in this way are fully exposed to the light and can intensify the processes of photosynthesis.*

21 *During the night, it stretches out its tentacles, armed with large numbers of nematocysts, to catch plankton.*

22 *The Indo-Pacific thick-stemmed Umbellate coral* (Lobophyllia) *has large corallites and thick, fleshy polyps, and is capable of ingesting quite large organisms as food. (Red Sea)*

23 *Mature Mushroom corals* (Fungia) *sometimes lie in dense concentrations on the seabed. (Off East Africa, Indian Ocean)*

24 *Stormy seas raging across the seabed also destroy corals. Here an old colony of Umbellate coral (Lobophyllia) is being broken up. (Gulf of Aqaba, Red Sea)*

25 *Tidal waves in open seas have tremendous force. More than 250 years ago, a frigate was destroyed out on the reef, but this cannon-barrel weighing about a ton was thrown up on the shore. The embankment behind it consists of coral rubble that has been cast up, and is several hundred metres wide. (Takapoto Atoll, Tuamotus, Pacific)*

26 *It is not unusual for coral growing near the surface of water, such as these finely-branching Staghorn corals (Acropora aspera), to be exposed to sunlight for an hour or two at low tide without suffering any harm. (Great Barrier Reef, Australia)*

27 *Compact colonies of stony corals close to the surface of the sea dry out at low tide, sedimentation causes the central part to die and only the edges continue to grow. Here, Wreath coral (Goniastrea pectinata) forms a micro-atoll. (Red Sea)*

54/55 28 | 29

28 *In areas of calm water such as lagoons, constant sedimentation causes part of coral colonies to die. Here, mussels have attached themselves to the dead skeleton of an Indo-Pacific Brain coral (Platygyra daedalea). (Tuamotu Archipelago, Pacific)*

29 *Until recently, the specific classification of the plank-like and the branching forms of Atlantic Fire corals was a subject of controversy. This colony shows that they belong to a single species, Millepora alcicornis, which produces the closed, plank-like form of growth in agitated water, and the branching form in the calmer water behind. Transitional stages between the two can be seen clearly. (Virgin Islands, Western Atlantic)*

56 30
31

30 *Information on the rate of growth of individual species of corals can be obtained by measuring those growing on a submerged wreck for which the date of sinking is known. (Red Sea)*

31 *On the outer slopes of reefs in the Central Pacific lying in shallow water, the principal corals represented are Claviform coral (Pocillopora) and table-like colonies of Staghorn coral (Acropora). (Society Islands, Pacific)*

and the Sea Trumpet *(Charonia)* have sizeable, strong-walled shells. When the animals are dead, the shells are cast up onto the beach, become part of the structure of the reef or accumulate in deep water at the bottom of the reef's outer slope, where they pile up in hollows to form lifeless, closed associations.

Less conspicuous are certain limestone-shelled, unicellular organisms that occur in large numbers in the form of sinking algae (Peridiniaceae) or as foraminifers (Foraminifera) that live in the reef (Fig. 18,1). Although scarcely visible to the naked eye, their contribution to reef construction is not inconsiderable. The Indo-Pacific foraminiferan, *Marginopora vertebralis,* is an exception, being large enough to be seen without difficulty. *Marginopora* also lives in symbiotic association with zooxanthellae. They are important in consolidating the reef; on the one hand they prevent sand drift by binding loose particles with pseudopodia that they extend through pores in their limestone shells in order to obtain food; on the other hand, with the help of zooxanthellae, they are able, like corals, to convert inorganic carbon in solution into bound organic carbon and calcium carbonate in a process of photosynthesis. Another striking foraminifer is the bright red *Homometra rubra.* It lives as a sessile organism on dead coral substrate. Massive occurrence of foraminifers is said to have a local effect on reef construction greater than that of corals. But measured against the long extinct foraminifers that grew up to 15 cm long, and the nummulitic reefs built by them in the Permian, their contribution to the biogenesis of recent reefs is small.

Echinoderms (Echinodermata) are curious animals, which—with the exception of the Sea Cucumbers—exhibit a radial symmetry of form that dispenses with the normally accepted structural elements of a front end and a back end. The calcareous defensive plating and spines (Fig. 19,3), which after the crea-

ture's death are incorporated into the reef, emphasize the impression of rigid and primitive organisms, and indeed they already inhabited the oceans in the Cambrian Period more than 500 million years ago. Geologically the most ancient are the Sea Lilies (Crinoidea) that show many primitive features in structure and habit. Whereas the Sea Lilies in the depths of the ocean raise themselves on long fragile limestone stalks from the dark sea bed where oxygen is scarce, those in the reefs are unstalked and free-swimming feather stars, leaving their refuges in the dusk of evening and moving on fine spindly limestone tendrils to seek out a familiar place at a rather higher level at which to catch their nightly supply of plankton (Ill. 43, 49, 50). To this end, they unfurl the feathery arms that have earned them their name. Movement of the arms permits a degree of swimming. Since almost the entire surface of the body is protected by limestone coverings, they have few enemies, apart from certain parasites and commensals.

Very similar in outward appearance to the Sea Lilies are the slender Brittlestars (Ophiuroidea), but they can be distinguished easily by the absence of filaments on the underside of the oral disc. And although the arms are furnished with spines, they are never feathery. The diver will find them almost exclusively in dark, sheltered places such as beneath stones and corals. When disturbed, they make themselves conspicuous by their lively movements as they scuttle for cover in shadow or among the branches of a coral. Basket Stars *(Gorgonocephalus)* hold a special place, with their branching arms that, at rest, are rolled up into an inextricable tangle, but which can extend to a length of 70 cm.

Most Starfishes or Sea Stars (Asteroidea) have five arms, others are multi-armed or the arms have regressed (Ill. 50, 124). The calcareous skeleton is composed of coarse bony bars, plates, spines and discs, providing a fairly rigid protective covering. Starfishes, sea urchins and sea cucumbers have a mode of locomotion unique in the animal kingdom: the ambulacral system. From a central ring canal, radially-branching water channels course into each arm. Branches arising from the radial canals terminate externally as hundreds of podia or tube-feet arranged in rows, protruding through small holes in the skeleton. When water is forced into them, the suction discs bulge outwards, the feet lose contact with the ground and make a coordinated movement forward in the direction of travel. As pressure is relaxed, they cling to the ground and draw the starfish along. Species without podia, such as the Burrowing Starfish *(Astropecten)* live on sandy ground in the reef or lagoon.

Sea Urchins (Echinoidea) have a shell or test of closely fitted plates. Species living on rocks exhibit globular radial symmetry, those that progress over soft ground a somewhat elongated, discoidal bilateral symmetry. The former have large, well-developed spines that move in ball-and-socket joints (Ill. 53, 54). In the Diadematidae (Longspined or Hatpin urchins) they take the form of long hollow needles, in Cidaroidea and *Heterocentrotus* (Slatepen urchins), of massive bars, round in section, or of angular clubs, in *Podophora* that lives in the surf region, of plates. Between the spines are large numbers of pedicellariae, stalked gripping claws that are usually three-jawed. Depending upon their shape, these small organs are used to secure food, to clean the test and the spine covering and, since some are equipped with poison glands, to ward off

enemies. The complicated toothed feeding apparatus has been given the name "Aristotle's lantern" on account of its shape and in recognition of the man who discovered it. The ambulacral feet are differentiated in form and fulfil a variety of functions, as mouth appendages, sense organs, "breathing" organs or suckers used for movement. Locomotion is assisted by spines situated on the underside, or may be effected entirely by them.

Sea Cucumbers (Holothuroidea) have an elongate body and are rather like fat worms. Calcareous skeletal elements in the form of ossicles or spicules that can be anchor-shaped, plate- or wheel-shaped occur within the muscular body tube, giving the creatures a membraneous to stiff, leathery appearance. Compact species *(Stichopus)* grow to a length of 1 m, tubelike species *(Synapta)* to more than 2 m. Since they are typically sand-eaters, most of them live on sandy ground. Sea Gherkins *(Cucumaria)* have a ring of well-developed retractable branching tentacles that frame the mouth. Since the podia allow them to climb among rocks, they seek out positions in the reef that offer a good supply of plankton. Paired, highly-branching tubules known as respiratory trees lead into the cloaca, through which water is pumped in order to provide the body with a constant supply of oxygen.

This wealth of organisms of the greatest diversity all produce calcium carbonate. But although one group or another may sometimes achieve local predominance, it is nevertheless the stony corals that make the greatest contribution to reef building today.

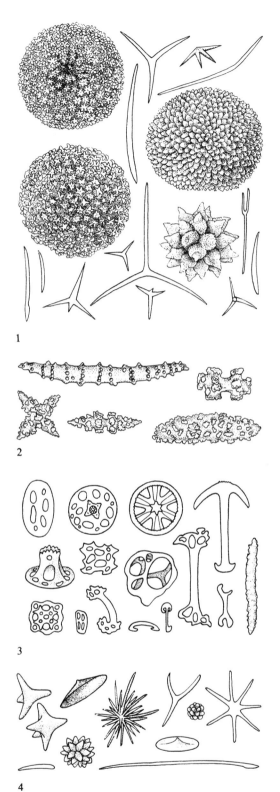

1

2

3

4

19 *Skeletal spicules, granules and plates in the reef sediment come from different creatures, 1 sponges, 2 gorgonians and soft corals, 3 sea cucumbers, 4 sea squirts (After various authors)*

Coral mountains

in warm tropical seas

Man, who has always striven to comprehend the images of his environment and to penetrate the world with his intellect, finds himself in the course of this process constantly prevailed upon to define terms and make classifications in order to achieve an overall view and at the same time to make clear intricate causal associations. In the sphere of biology, apparent analogies have frequently led to wrong conclusions. One need consider only how, because of superficial similarities, the whale was at one time classified as a fish. Corals fared no better—first they were plants, then stones and finally animals.

While coral reefs were still being discovered by early travellers, they were classified into typical forms, and the terms "lagoon islands" or "atolls", "barrier" or "canal reefs", "fringing" or "littoral reefs" came into use. Although in many cases still appropriate, this classification is no longer adequate, particularly since the terms are based only on external characteristics, and tell us nothing about the origins of the individual reefs or their ecological situation.

Coral reefs are subject to processes of evolution and ageing. Darwin was one of the first to point out that it is rare for them to persist in an unaltered state for a long period, and here he was undoubtedly thinking of geological periods. This constant change led, in many cases, to the view that reef types and reef structures differ so greatly from locality to locality that they defy classification. But this is not so. Rather they developed as the result of particular geological and hydrological situations that recur in very different parts of the ocean, just as the types of reef occur repeatedly.

Criteria for the classification of coral reefs, apart from shape and size, are their location in shallow or deep water, on the lee or weather side, the history of their development and the principal organisms responsible for building them. All these factors bring about particular characteristics, for example, designation of location implies certain ecological conditions. Particularly in view of the dangers that threaten coral reefs as a result of human action, it seems appropriate to give more consideration to the ecological aspect than has been done hitherto.

Linear and circular coral reefs

Since hermatypic corals can achieve full vigour of growth only in warm clear sea water suffused with light, reefs are found everywhere in the tropical littoral where no cold upwelling currents or sedimental river inflows alter the hydrological picture. Belts of coral reefs, often hundreds of kilometres long, lie off these coasts. Because of their elongated form stretching as a continuous line parallel to the coast, they are classified as linear reefs. In contrast there are those reefs with the outline shape of larger or smaller patches. They are described as circular reefs. Within these major groups are various types of coral reef (Fig. 20), with numerous transitional stages that embody all phases of development. Associated with them are various coralline structures—not true reefs that have grown upwards from the sea bed, but diffuse coral grounds or depressions in the sea bed where corals predominate. In regions with especially prolific development of coral reef complexes, such as the Great Barrier Reef or the Red Sea, several types of reef are found.

Linear coral reefs extend parallel to the coast and their highest parts lie just below the surface of the water. They are exposed to the action of the waves that surge in from the open sea and break upon them, and therefore exhibit a distinct weather or windward side (the reef front). A linear coral reef always grows outwards into the ocean. It is possible to distinguish three principal forms of linear reef: fringing reef, bank reef and barrier reef.

The fringing reef is common in all tropical seas. Growing directly from the shore line, it fringes the coast, sometimes for a distance of many kilometres (Ill. 64). Its height coincides with the low water mark. The outer slope and the reef crest are often marked by small pothole depressions, and in tidal seas, by shallow, narrow grooves running at right angles to the shore (Fig. 22). As it increases in age, the outer edge is pushed

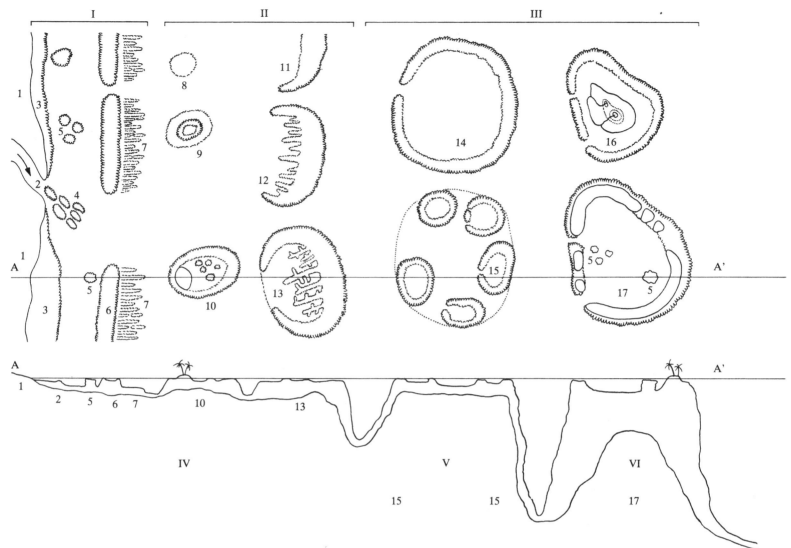

progressively further forwards as a result of the particularly vigorous growth of corals in turbulent waters, and the width of the reef crest can reach several hundred metres. Since the water in the reef flat behind the outer margin is usually very calm, and sedimentation can be considerable, coral growth towards the land decreases. Parts of the reef flat can be exposed. The depth of the outer slope depends upon the inclination of the substratum and can be over 20 m. The more acute the seaward slope of the coast, the narrower the fringing reef. Off shore, there is often a shoreline channel in the form of a narrow depression.

Another linear reef structure occurs a few hundred metres out from the shore

from which it is separated by a lagoon 1–3 m deep and up to about 1,000 m wide. This reef, originally designated a "lagoon fringing reef", represents a separate type, the bank reef (Ill. 65). It is characterized by its shallow lagoon and its location in the midst of the coastal plateau. The outer slope rarely falls away to a depth greater than 20 m. Processes of erosion caused by the pounding of the waves and the barrier effect of water, combined with moving sand and pebbles, have driven indentations into the bank reef close to one another, producing a zone of grooves and spurs. Where the breakthrough has been completed and extended, reef canals separate discrete tracts of reef. Through them, a great deal

20 Diagram showing common forms of reefs and their distribution in the ocean. I Littoral reef area; II neritic reef area; III oceanic reef area; IV shelf base; V mountain base; VI island nucleus.
1 coast, 2 mouth of stream, 3 fringing reef, 4 quader reef, 5 horst reef, 6 bank reef, 7 reef tongues, 8 hill reef, 9 wreath reef, 10 platform reef, 11–13 Barrier Reefs (with 11 cuspate reef, 12 prong reef, 13 mesh reef), 14 major faro, 15 daughter faros, 16 rampart reef, 17 atoll. A–A' indicates the line of section. (Original)

of the water exchange between the open sea and the lagoon is accomplished.

Similar to bank reefs in their development and dimensions are bar reefs, except that they lie across the curve of a bay or inlet, cutting it off from the open sea. The bay thus becomes a basin with relatively calm water. If the bay is a wide one, and the seaward slope only slight, the bar reef extends broadly, blurring the typically compact, clearly articulated structure of the bank reef by a diffuse scattering of reef components.

The most imposing of the linear reefs are the barrier reefs (Ill. 67). They grow up from a base of postglacially submerged mountain ranges. Yet in the Great Barrier Reef they are no more than 22 to 27 m thick, in the Red Sea about 20 m. The outer slopes merge into the rock faces of the mountain and fall away steeply into the depths of the sea. At 50–70 m down, they are still covered with hermatypic corals, so that the boundary between primitive rock that provides the foundation and the limestone of the reef is masked by coral encrustations. Barrier reefs lie some considerable distance off a coast. Their length varies. Off the coast of Sudan, there is a barrier reef 330 km long. New Caledonia has a barrier reef of more than 700 km length. The Great Barrier Reef extends for more than 2,000 km at a distance of 30 to 150 km off the east coast of Australia. It is the largest coral reef in the world. All the lagoons are broad and deep; those of the Great Barrier Reef, with an average depth of about 40 m, almost equal the dimensions of the Baltic, while those of the Sudanese barrier reef reach a depth

21 Development of a wreath reef.
1 Stony corals establish themselves on a favourable part of the rocky sea bed, grow, reproduce and eventually form a reef.
2 If the depth of the water above the reef increases, the reef grows upwards: a hill reef has been formed.
3 Inside, where water movement is slight, the corals die, accumulated water depresses the reef surface, a wreath reef is formed. (Original)

of more than 400 m. Circular reefs and islands may occur locally in the lagoons. Reef passages divide a barrier reef into separate sections. Turbulent masses of water forced through by tides and winds encourage particularly vigorous growth in the corals colonizing the margins, and this growth is continued shoreward at the sides. As a result, the reefs eventually take on an apron-like shape, curving on the leeward side, or other forms depending upon the stage of development (Fig. 20, 11–13). The breadth of reef

crests usually varies between 100 m and 1,000 m.

Circular coral reefs can be patchy, roundish, elongated oval or annular. They are in a state of constant change. It is characteristic of them to have coral growth all round or at least on two-thirds of the outer face. Overall, circular reefs make up a very heterogeneous group. Frequently they lie in extensive lagoons and in sheltered bays. Prevailing air currents, such as the trade winds, determine the orientation of growth. The

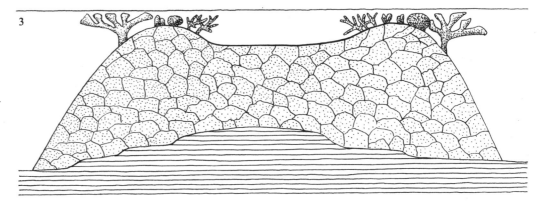

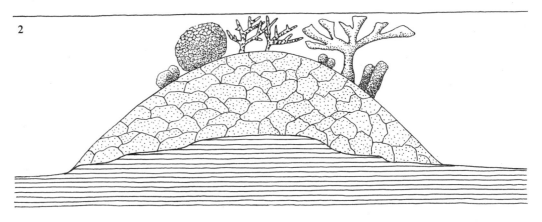

diameter can vary from a few decimetres to more than a hundred kilometres. Circular reefs include a number of forms, of which hill reefs, wreath reefs, platform, horst and dish reefs occur in shallow water, rampart reefs and atolls in deep water.

Hill, wreath and platform reefs form a developmental chain (Fig. 21). Hill reefs are like domes, the centre of which rises almost to the surface of the water. If the hill reef developed in turbulent water, it is oval and its orientation is typically determined by the prevailing direction of the waves; the windward slope that faces the waves is shallower and longer than the leeward slope, since there, a more prolific growth of coral has caused a greater production and deposit of calcium carbonate and thereby a greater extension of the reef. If a hill reef lies in relatively quiet water, it may be almost perfectly circular. The development of hill reefs is remarkable in that they can grow on sandy ground. Among the species of coral that are able to colonize both rocky and sandy ground, there are some widely distributed species such as Pore corals. On sandy ground in quiet waters they develop colonies up to 4 or 5 m in height and of corresponding diameter. On dead parts of these colonies, other species of coral settle and grow, from time to time being broken away by the buffeting of violent seas. They accumulate where they fall as a ring-shaped deposit of rubble, some of it still living, proliferate and over the course of centuries, add to the massive original nucleus, fusing together to form a hill reef on which various species of stony coral continually settle. In the same way, any irregularity on the hard ground can initiate the development of a hill reef.

In the further course of events, the volume and height of the hill reef increases to such an extent that the centre becomes subject to restriction in the circulation of water, as a result of

coral growth on the outer margin. The central zone is deprived of an adequate supply of fresh water, and rubble, gravel and sand are deposited there. Coral growth declines, while at the outer margin, a circlet of growth continues. Such a reef is called a wreath reef. Since it lies beneath the surface of the water and is relatively small, it possesses no independent community of calm-water organisms.

As the next stage in development, the platform reef is formed. While hill and wreath reefs are relatively short-lived, being in the nature of germinal or embryonic stages, the platform reef has a more permanent nature. As a result of continual accretion, a reef massif several hundreds of metres to several kilometres in diameter develops over thousands of years. Its margins have long since reached the surface of the water. Very often, storms have cast up boulders and coral rubble onto its edges so that a boulder zone rises above the ocean. Typically, aggradation on the leewards side causes the formation of islands (Ill. 69). The central depression is deepened by subsidence that is intensified by the weight of water standing for long periods at low tide. Scattered across it are coralline structures, systems of channels and sandy patches, and it may be crossed on foot at low tide. In tidal regions, where the weather side is exposed to the violent action of the waves, calcareous algae are the principal products of the reef mass. In front of the island, a reef plateau often extends to a steep outer slope that consists of branching and massive colonies.

The "faros" of the Maldive and Laccadive Islands have an atoll-like form (Ill. 63). As a result of peripheral growth on a huge underwater shelf at a time when the water level rose following the last glacial epoch, gigantic ring-shaped coral structures several hundred kilometres in diameter were produced. With the slight subsidence of the substratum, sea

erosion caused these formations to split into separate reefs and islands. The gaps torn in the continuous reef structure permitted a constant exchange of water, and for a second time, ring-shaped reefs and islands were formed with open water on all sides and where growth was predominantly outwards into the ocean. So the faros, also known as "shelf atolls", are not true atolls, but represent a particular, circular type of reef that has developed from huge platform reefs.

The pattern of development of horst and bowl or cup reefs is quite different. Horst reefs are the residual massifs of underwater terraces or reefs that have been eroded on their margins by the sea. Therefore they occur frequently as rows or groups. These limestone rocks lying in the shelf region show a crest close to the surface of the water and steep, fissured side walls, while the seaward faces are dissected by caves, channels and vents. Grooves have been cut into the floor by the action of water and gravel. The crest of the reef and the side walls are covered irregularly by coral growth. Where horst reefs are situated in protected lagoons and bays exposed to only slight winds, they are not differentiated into windward and leeward sides.

The bowl reef can develop from a horst reef that has gradually come to lie deeper under the surface of the sea, and in which increased growth of coral at its edge has caused their elevation. The higher the rim, the more the central hollow resembles a bowl. The floor of the bowl is characterized by surface grooves and channels (Fig. 23).

Rampart reefs and atolls lying in deep water represent a third line of development for circular reefs (Ill. 66, 68, 81). They are linked to island formation and their occurrence is usually associated with underwater volcanic activity. By the elevation of volcanic cones and mountains above sea level, particularly in the Pacific Ocean, whole archipelagos such as the Caroline and Marshall Islands,

22 Edge of a fringing reef lying in shallow water in Gulf of Aqaba with dense growth on slope of branching stony coral, particularly Millepora dichotoma *(Fire coral),* Porites *(Pore coral),* Acropora *(Staghorn coral),* Pocillopora *(Claviform coral) and* Stylophora *(Styloid coral). (After Mergner and Schuhmacher, 1974)*

the Tuamotu Archipelago and the Society Islands have been brought into existence. First the volcanic islands are surrounded by a fringing reef. Since their emergence is usually followed by a period of slow subsidence, and later by a post-glacial rise in the level of the oceans, the corals on the outer margin of the fringing reef continuously grow upwards to keep pace with the decrease in height. But behind the reef crest there is stagnation because as the outer rim grows higher, the inflow of fresh sea water to the inner part of the reef is reduced. The resulting increase in sedimentation finally prevents growth in the rearward parts of the reef entirely. In the course of further subsidence, the original fringing reef has developed into a rampart reef (Fig. 26) that is separated by a

lagoon from the island core. The lagoon is frequently less than 1 km wide, but currents have scoured out a deep natural channel along its centre. Where rivers flow into the ocean from the island's mountains that can be 1,000 m or more in height, reef growth has been interrupted by the quantities of silt carried along in their waters. At these points, there are deep breaches in the rampart reefs, and currents flow into the lagoon from the sea. Now corals can grow again on the inner slope of the reef, and once again a fringing reef can develop off the shores of the island. The inner slope of the rampart reef, highly articulated and covered with coral growth, now merges gently into the lagoon, but at the outer slope, beyond a terrace lying at a depth of 15 to 30 m, the side slopes down.

At a further stage of development, the nucleus of the island is again under water. The rampart reef has grown progressively higher, forming a ring-shaped reef that reaches up to the surface of the sea; it is known as an atoll. The outer slopes that fall away steeply into the depths of the sea from a narrow reef ridge are truly magnificent, adorned with a dense growth of coral. Stony corals are more prolific here than on other reefs, possibly because there is a complete absence of inflowing terrigenous elements such as fresh water and sedimentary matter.

The outer reef ring towards the open sea is much more exposed to wave action on the windward side than on the opposite leeward side. Between the two is an area that might be called the semi-leeward zone. The sea bed that falls away steeply to a great depth allows the waves to flow past the leeward side of the atoll unhindered, evenly and with considerable stability of direction. On shallow ground, such as is frequently found round platform reefs and faros, the waves come into contact with the ground as they wash round the reef.

The larger the submerged island, the greater the diameter of the atoll, which in some cases is more than one hundred kilometres (Fig. 24). Since even here, the lagoon water cannot flow away sufficiently quickly as the tide ebbs, a barrier of water accumulates, the weight of which—and it may be many tons—increases in proportion to the diameter of the lagoon. As a result, the sedimental base that has extensive cavity formation, is compressed and deepened. The same principle doubtless applies to the lagoons of other types of reef. Depending upon the diameter of the atoll, the depth of the lagoon varies from 30 to 80 m. Within the lagoon, scattered hill and wreath reefs can be found, densely covered with a growth of corals and shells. Alluvial accretion leads to the formation of islands on parts of the reef ring, and the

characteristic landscape that develops corresponds to the Maldivan word "atoll", signifying islands lying in a ring. Until the foundation of the Maldivan Republic, each governmental district of the Sultanate was termed an *atolu*, the local governor the *atoluveri*. Even seen from a distance, atolls lying in the middle of the ocean present a picture of unforgettable beauty. Breakers crash foaming and thundering against the outer edge of the reef. Behind them stretch miles of pale sand and pebble shore, crowned by a grove of coconut palms spreading cool shade and swayed by trade winds. This ring of atolls encloses a quiet lagoon, which is like a peaceful oasis in the midst of the turbulent ocean.

In the course of geological history, many reefs have been elevated above the water (Ill. 70), others have been overlain by rubble and alluvial sand. In both cases, coral islands have developed. Currents drive plants, algae and dead organisms onto the shore, which decompose and provide humus for the infertile sandy soil. Seeds of plants that thrive in conditions of drought and salinity are carried ashore by waves. Some of them germinate to become the first, meagre vegetation. Soon coconut and pandanus palms, bushes of *Messerschmidia* and *Suriana*, later on Casuarina trees, *Pisonia* and breadfruit trees spread their branches. Their foliage provides food for insects washed up on pieces of wood or carried along in air currents. Lizards feed on the insects. Hermit crabs *(Coenobita)* and land crabs *(Ocypode)* inhabit the foaming water's edge. Birds begin to nest. Man and rats are the next arrivals. Heavy monsoon rains provide drinking water for all, and man soon finds out to collect the rain-water in large shells of giant clams. Coconut palms supply food, clothing and building materials. Fish, mussels and lobsters are caught in reef and lagoon. A coral island has been established (Ill. 74–80).

65 32

32 *Red Grouper* (Cephalopholis aurantius) *and some Jewelfishes* (Anthias squamipinnis) *are on the alert for enemies. (Red Sea)*

66/67 33 34 | 38 39
35 | 40
36 37 | 41 42

33 *Gorgonian coral, such as this* Eunicea laxispica, *prefers to settle on rocky projections, since from here it can most easily benefit from plankton carried past in water currents. (Virgin Islands, Western Atlantic)*

34 *Luxuriant colonies of corals such as green* Stylophora pistillata, *brown* Montipora *sp. and violet* Distichopora violacea *grow on the sides of trenches and channels at a depth of 1 to 3 m, where there is a limited but active exchange of water. (Red Sea)*

35 *The glowing red of the Gorgonian* Paramuricea clavata *is visually ineffective at a depth of 15–20 m where it grows. (France, Mediterranean)*

36 Paracirrhites arcuatus, *known as "Coral Watchman", constantly perches on branches of coral on the lookout for prey. (Society Islands, Pacific)*

37 *Where no corals grow, the rocks are covered by various species of calcareous algae, mainly red. (Society Islands, Pacific)*

38 *Small and great Button corals* (Montastrea annularis) *occur frequently in communities in the Western Atlantic in depths of 5–40 m. (St. Croix, Virgin Islands)*

39 *Delicate Filigree corals* (Stylaster elegans) *grow in darkness in the shelter of small caves. (Micronesia, Pacific)*

40 *At a depth of 20 m, flat* Mycetophyllia danaana *and* Agaricia agaricites *grow on a steeply-sloping wall. (Virgin Islands, Western Atlantic)*

41 *Branch of a Gorgonian coral* (Eunicea sp.) *with extended polyps. (Curaçao, Caribbean Sea)*

42 *Two males of the timid species of Single-spotted Butterflyfishes* (Chaetodon unimaculatus) *have just concluded a duel by threat, in which each keeps in sight the eye spot of the other. The fish indicating submission has reduced the spot and is retreating backwards from its rival. (Society Islands, Pacific)*

68 43 |

43 *Feather stars, that are active often only at night, remain in hiding under soft corals—a* Sarcophyton *above, a* Lithophyton *below. (Red Sea)*

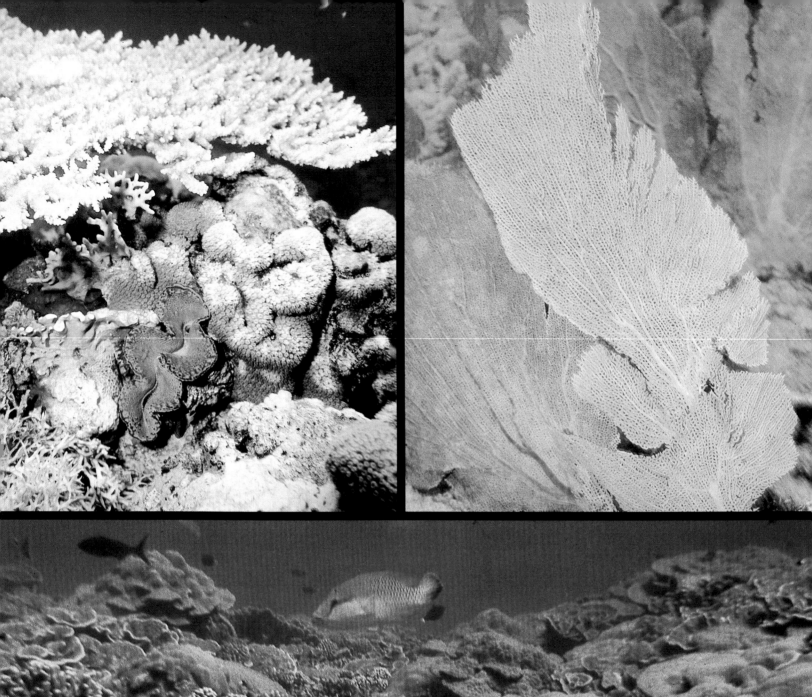

44   *The Hermit crab (Dardanus sp.) shares the
exterior of its snail-shell with a sea anemone which
keeps the hermit's enemies at bay with its stinging
tentacles, while at the same time benefitting from
the crab's leavings–a true symbiosis. (Red Sea)*

70/71   45   46 | 48
47

45   *At a depth of 10 m, clear water, gently inclined
slopes and constant gentle agitation of the water
provide excellent conditions for a varied coral
association. Here, Staghorn corals (Acropora
squarrosa), Fire corals (Millepora dichotoma),
Scalpula corals (Galaxea fascicularis) and Needle
corals (Seriatopora hystrix) have colonized an area
round about a Giant Tridacna (Tridacna maxima).
(Red Sea)*

46   *Sea fans (Gorgonia sp.) occur in two colours,
yellow and reddish-violet. They are flexible and
tough, often growing in surf. (Virgin Islands,
Western Atlantic)*

47   *Pore corals (Porites), Micropore corals
(Montipora) and Claviform corals (Pocillopora)
often grow in dense associations at considerable
depths in Pacific reefs. (Tuamotu Archipelago,
Pacific)*

48   *A delicate red Horny coral (Melithaea sp.)
hangs from the corner of the roof of a cave, its white
polyps show an effective contrast to the background
of black water. (Red Sea)*

72   49

49   *In its active evening phase, the Feather Star
Dichrometra sp. has taken up the accustomed
position from which it catches plankton. (Red Sea)*

Further classifying features and types of reef

The characteristics by which a coral reef is classified are visible externally. Then they are named. The hydrography of the region defines the particular ecological situation. Stony corals respond to a slight change in salinity caused by fresh water, with a reduction in growth. In addition, littoral reefs are subject to an inflow of terrigenous materials from rivers, rainwater run-off and high sedimentation—all negative factors for stony corals. Reefs lying off coasts can undoubtedly show good development, but anyone familiar with the reef situated far out in the ocean will appreciate the difference between the good condition of the former and the prolific luxuriance of the latter. The quality of the surrounding water is thus of prime importance in determining the qualitative and quantitative differences in coral growth. It is therefore appropriate to regard the description of the reef position as a component of the categorization of reef type.

Reefs lying off coasts are called littoral reefs (Fig. 20). They include reefs such as fringing, bank and horst reefs situated on a shelf off continents and large islands, where they can be reached by terrigenous sediments. The neritic coral reef lying in open sea is no longer within range of terrigenous silt, but is nevertheless affected by terrigenous matter in solution, such as humic acid, that is carried into the sea by fresh water. This category includes barrier and platform reefs lying moderately close to land and rampart reefs that surround islands. Oceanic reefs are those that arise in the middle of oceans, far from any terrigenous shores, as faros or atolls. Land-derived elements reach them only as minute traces, such as are distributed throughout the entire ocean. They do not lie on a shelf. It should be noted that a reef can be influenced partly by the one and partly by the other pattern of circumstances. For example, the coastward side of a rampart reef can border on a neritic lagoon while the seaward side that slopes steeply down to the deep sea is washed by oceanic waters. The situation is similar in many barrier reefs and other reefs situated on the edge of a shelf. Since sea water is an unevenly-flowing medium, the transitions are also fluid.

Reef-building by corals is not always a continuous process. Not infrequently, submerged limestone terraces and plateaus are eroded to such an extent that only elevated residual structures are left. These limestone bases that only subsequently are covered by coral growth, can, as secondary reefs, be distinguished from primary reefs that were built up directly by corals.

Reefs can be classified in greater detail in terms of the principal organisms that constructed them. They can be called *Porites* reefs or *Acropora* reefs, where the contribution to the reef by these particular species has been especially great. In the Mediterranean off Greece, at a depth of 18 m, coral banks 1–10 m long and up to 8 m high have been discovered, that were formed by the coral *Cladocora cespitosa*. On them, sponges, tube worms, hydrozoa and the bryozoan *Retepora* have established themselves. Nudibranchs are also common there. Damselfishes *(Chromis),* sea bass and seabream are constantly to be found in the vicinity of these *Cladocora* banks, the ecological independence of which is beyond dispute. An unusual coral association is found in the North Atlantic along the edge of the European continental shelf at depths of between 60 and 2,000 m, but particularly frequently between 130 and 400 m. Preferably in areas of deep-ocean currents, the para-symbiotic, bush-shaped *Lophelia pertusa* (Linné) forms extensive hedges. Even

though it builds no reefs, it nevertheless creates a biotope of its own, where it is joined by many other organisms to produce a characteristic biocoenosis.

Lophelia banks have also been observed in other regions of the ocean where conditions are similar. Locally, certain reef builders can outdo the Scleractinia in the production of limestone, and there, *Marginopora* reefs come into being, or where both contribute about equally, a *Marginopora*-Scleractinia reef.

Along the rocky shores of the Mediterranean, bracket-shaped projections about a metre thick are sometimes found, that have been produced by calcareous algae. True reefs built by lime-depositing algae (see p. 44) can be found in all oceans, particularly off islands exposed to heavy seas, for example, in the Atlantic off the island of Fernando da Noronha that lies

350 km from the coast of Brazil, off the Cape Verde Islands and off the Virgin Islands. In the region of the trade winds, representatives of the genera *Porolithon, Goniolithon, Lithothamnion* and *Melobesia* in particular deposit thick, hard, very resistant limestone encrustations. They cover dead corals with these secretions that bind together coralline rubble. Off the coast of Venezuela, lime-depositing algae have formed sinter-like off-shore terraces. Off the western coast of Australia is the largest recent Stromatolite reef in the form of petrified circular boulders, columns and ridges (Ill. 72). Masses of blue-green algae proliferate in a hyper-saline bay. Calcareous sand and mussel shells become entangled among their filaments and are cemented into stromatolites by the precipitation of calcium carbonate. Since the rate of growth

23 Sketch of a bowl or dish reef in the Bermudas, that has been built primarily by calcareous algae. (After Moore from Shinn, 1971)

is extremely slow at 0.5 mm per year, they are undoubtedly thousands of years old. Stromatolite reefs were formed more than 600 million years ago. They represent the earliest biogenic reef formations on earth, and at that time were common and considerably larger than they are today.

Bryozoa (see p. 46) are important reef-building agents in some reefs, such as those of Bermuda. The opportunist species *Membranipora reticulum* L. that has wide distribution, develops such rank growth in brackish waters that it forms small reefs. In they Bay of Kerch in the Black Sea and in the lagoons of Comacchio in the Adriatic, they form colonies up to 2 m in diameter—reeflets rather then reefs, it is true, particularly when compared with the impressive fossil bryozoan reefs created by their ancestors millions of years ago.

Vermetids (worm shells) are outsiders in that they secrete coiled shells and move about freely only when they are young. Later, the locomotory foot atrophies and the shells are extended to form irregularly coiled limestone tubes that are cemented firmly to an available base by limestone secretions. Individually, they help to consolidate the reef structure. When they occur massively, such as along coastal strips and round islands of Brazil, they create Vermetid reefs. Over a period of 3,000 years, *Vermetus nigricans* constructed a reef barrier 35 km long off the Florida archipelago. Incrustations of *Dendropoma* have been described on the Mediterranean coast of Israel and in the Gulf of Aqaba. Ringed worms that live in limestone tubes are also capable of forming small reefs when they appear in vast numbers (Ill. 70), as serpulids have done in Baffin Bay. In the Bermudas, Serpulid reefs exist with a diameter of up to 30 m.

The upright chitinous tubes in which *Phragmatopoma lapidosa* live, combine with the sand-encrusted tubes of *Sabellaria floridensis* and *S. vulgaris* to form Sabellarian reefs one metre high. They are found in very shallow water at the edges of the beach where they are exposed at low tide. In some places, they extend for miles along the Atlantic coast of Florida. The tubes of these small worms, that are only a few centimetres in length, calcify in the lower part and become rigid. Off the north Atlantic coast, *Sabellaria spinulosa,* known as "sand coral", builds similar reefs. But calcification does not occur, and the reefs do not endure for long. Off the North Sea island of Norderney, 75 million individuals created a Sabellarian reef 8 m wide, 60 m long and 60 cm high. Then it was destroyed. Observations of a similar phenomenon off the English coast have been reported from Duckpool.

All biogenic reefs, whatever their nature, are extremely complicated structures. Usually, many different organisms contribute to their growth. In large reef tracts, there are almost always several different types of reef. In such cases, classification of a single reef type would give a distorted picture. But this very complexity makes it essential to attempt a review based on the principle of causality of the coral reefs of the world.

Diversity of reef structure

The wide variety of factors that affect reef development have produced a wealth of different reef structures. In addition to the types of reef already described, there are examples of coral colonization that either represent very divergent reef forms or are not reefs at all in the sense of eminences rising from the sea bed. One of these curious reef forms is the block reef. It occurs where a river, fanning out broadly, flowed across a limestone terrace into the sea during the wet post-glacial period, and in doing so, cut out of the plateau erosion channels that diverged slightly towards the sea. As a

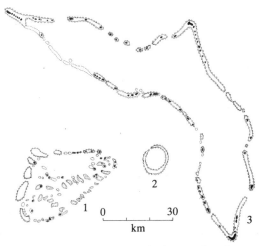

24 Different outlines shown by faros and atolls: South Mahlosmadulu, Maldive Islands (1), North Minerva (2), Kwajalein, Marshall Islands (3)

result of a slight rise in the level of the sea, water covered the limestone ridges that were mostly no more than a few metres wide. The sea forced its swirling waters through the system of channels, widening and deepening them. In places, pebbles stirred into violent agitation, broke through the limestone ridges laterally, splitting them into separate block-shaped rock ridges. Later, the block reef was colonized by corals. It is certainly to be included among secondary littoral coral reefs, but can be classified neither as a linear nor as a circular reef.

Certain regions of the tropical shelf exhibit none of the characteristics of a reef and yet possess dense coral growth. Irregular rock structures at a depth of 5 to 30 m, broken by projections, prominences and potholes, in which there is a continuous supply of turbulent water rich in plankton and oxygen, promote the development of densely colonized coral grounds that provide a suitable substratum for various coral associations, depending upon the nature and vigour of the hydrodynamic conditions. The margins of underwater terraces are favoured by stony corals, for here water turbulence is high and sedimentation is prevented. Coral terraces are especially typical of coasts that have been subject to elevation and subsidence.

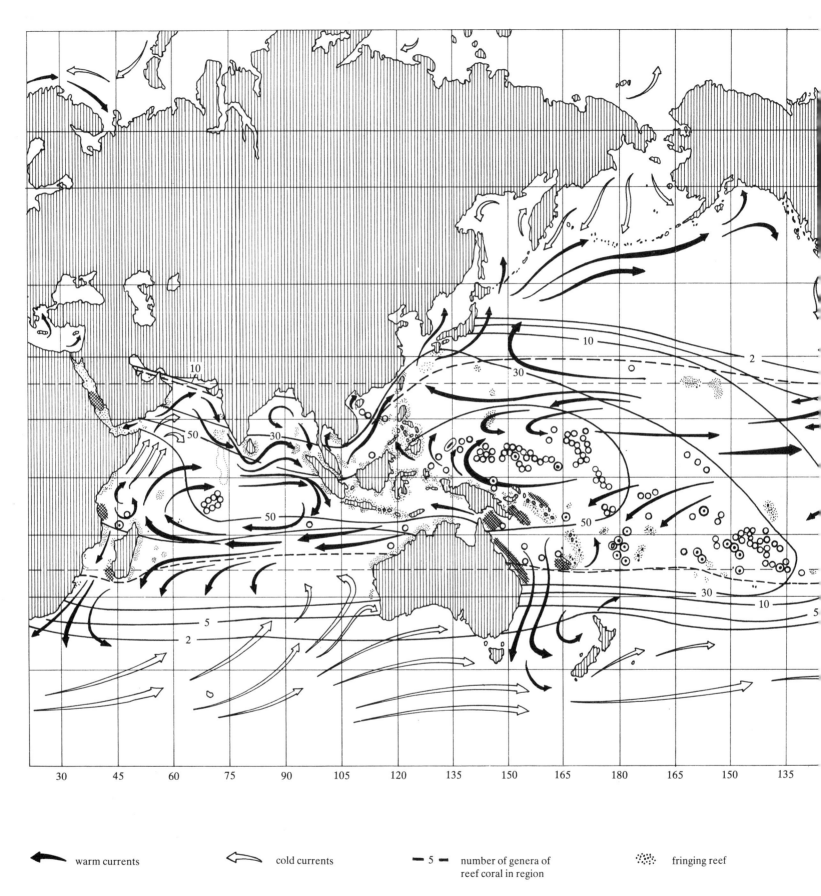

30   45   60   75   90   105   120   135   150   165   180   165   150   135

━━▶ warm currents   ⇦ cold currents   ━ 5 ━ number of genera of   ⋮⋮⋮ fringing reef
                                                            reef coral in region

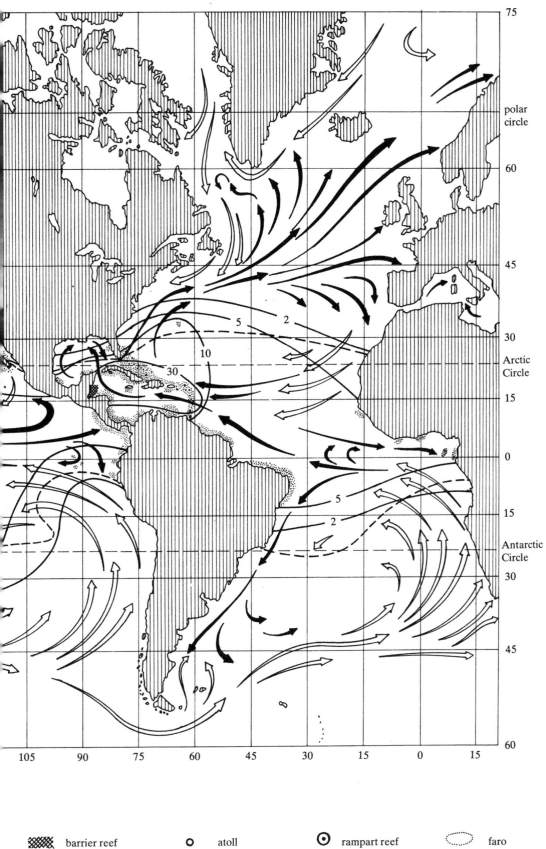

25 Geographical distribution of recent reef corals and coral reefs. Basically, the present geographical distribution of hermatypic corals follows the warm ocean currents and the 20°C winter isotherms. The figures refer to the number of the genera of stony corals of the area.
————20°C winter isotherms

75

polar
circle

60

45

30

Arctic
Circle

15

0

15

Antarctic
Circle

30

45

60

105   90   75   60   45   30   15   0   15

▨▨▨ barrier reef   O   atoll   ◉   rampart reef   ⬭   faro

The coral canyon is not unusual. It has arisen either as the result of subsidence at a river mouth or from the remnants of a reef canal. Water masses flowing in from the open sea are directed downwards with violent eddying action along its edges. In heavy seas, one surge of water after another forces its way through the gorge-like narrows. Here too, favourable water movement, very broken side walls and lack of sedimentation provide conditions of optimal growth for well-developed coral colonies.

Frequently, individual large species of coral develop striking shapes. They are always species that form compact, firm colonies that persist for centuries, Pore corals in particular. Growing at a depth of 10 m, the colony can build a tall, slender coral pinnacle that grows up to the surface of the water. The lower part is dead for the most part, and is colonized by other sessile organisms. But the top few metres are alive and identify clearly the agent responsible for the structure. Very similar to the coral pinnacles are the coral "bommies" of shallower waters. They are irregularly spherical in shape, mostly broader than tall, and up to 6 m in height. They too are frequently formed by *Porites* colonies.

Many heads, many theories

In the final phase of the Age of Discovery, when man turned his attention to a detailed examination of recently discovered continents and oceans in order to derive from them the greatest possible benefit, natural scientists also took part in extensive sea voyages. Johann Reinhold Forster and his son Georg Forster accompanied the Englishman James Cook, Chamisso the Russo-German Otto von Kotzebue, Darwin Captain Fitzroy. Although the purpose of these journeys was largely political and economical, aimed at extending the power of the

motherland, and the sailors looked upon the men of science as tiresome ballast rather than as useful members of the expeditions, the scientists themselves were so inspired by the wealth of new impressions that they willingly endured all inclemencies of weather, the privations and illnesses of a sea voyage in order to enlarge man's conception of the world. For all who travelled in the tropics, one of the foremost questions to occupy them was that of the origin and development of coral reefs.

The observations made by Georg Forster may seem somewhat ludicrous today, but he was, after all, writing more than 200 years ago. "The coming into being of these coral rocks presents us with no less an admirable example of the omnipotence of the Creator, who is so often able to achieve great and ultimate purposes through the slightest of means. As is generally known, coral is a structure built by a small worm which enlarges its house progressively as it itself grows. In this small creature, one can scarcely observe sentience sufficient to enable one to distinguish it in this purpose from the plants. Yet from the fathomless depths of the sea, it builds a rocky structure that extends up to the ocean's surface, rising to a height to provide a solid ground on which countless men can find a residence... The worms that construct the reef seem to have the impulse to protect their dwelling from the might of the wind and of the violent sea. Therefore they lay out their coral rocks in hot regions of the world, where the wind blows almost always from the same direction, in such a way that they form, as it were, a circular wall and separate off a lake from the remaining ocean, in which no frequent movement occurs and the polyp-like worms find a peaceful dwelling."

After Georg Forster's recognition of the importance of stony corals in reef construction, Alexander von Humboldt (1806) expressed the opinion that atolls

develop on the edge of old volcano craters, which however, can occur only if the edge of the crater lies close to the surface of the water. At best, such a situation could exist only by pure chance. Eschscholz and Chamisso suggested, more appositely, that massive corals prefer surf, reef margins therefore grow particularly rapidly on the outside and so form a ring.

In 1837, Charles Darwin put forward his now famous subsidence theory in the development of coral reefs to the rigorously critical committee of the Geological Society in London. Since it is only in the light-flooded zone near the surface of well-aerated, turbulent water that corals are able to grow strongly enough to produce the vast quantities of skeletal material required for reef formation, the fringing reef develops first of all on a slowly sinking rock littoral directly on the coast. With the further continuous subsidence of the land and of the sea bed, barrier or rampart reefs develop, and with the complete disappearance of island cores beneath the surface of the sea, finally atolls. Many decades after his theory of reef development had been put forward, drilling operations carried out on the Bikini atoll showed coralline rock still at a depth of more than 800 m, and in the Eniwetok atoll, the anticipated basalt group was reached only at 1,400 m. These two cases are direct evidence in favour of Darwin's subsidence theory. No doubt other reefs also developed in this way, and would, if examined, provide similar evidence of it. The subsidence theory thus explains the development of several major classes of coral reef. In terms of its ecological reasoning, it is empirically correct, though causally somewhat incomplete. With the increase in scientific knowledge, however, its "geological component" was seen to require revision. In Darwin's time, discussion of the effects of glaciation on the oceans was only just starting, and had not yet been widely accepted. He men-

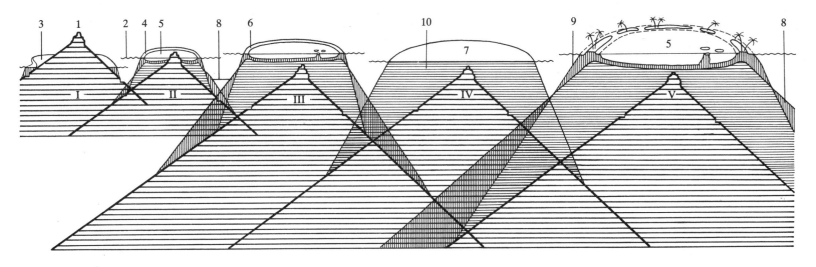

26　Development of a fringing reef, rampart reef and atoll according to the compensation theory.
Phase I: A fringing reef has developed round a rocky oceanic tropical island.—Phase II: With subsidence of the island and a rise in the water level, the fringing reef grows upwards and outwards to form a rampart reef.—Phase III: The antagonistic forces of rising water level and island subsidence lead to the complete submersion of the island, during which process, the corals compensate for the loss of height by constant upward growth, and an atoll develops.—Phase IV: The glacial binding of vast water masses as continental ice brings about a second lowering of the level of the oceans, as a consequence of which, the atoll is flattened and levelled by the action of moving water.—Phase V: Post-glacial melting raises the level of the oceans again, and the corals that colonize the stump, compensate for this rise by progressive upward growth, producing a recent atoll with a considerable increase in mass.
1 island or island core consisting of primary rock, 2 water surface, 3 fringing reef, 4 rampart reef, 5 lagoon, 6 atoll, 7 platform of the atoll stump, 8 slope of coral debris, 9 loosely-fused recent coral limestone and living coral growth, 10 older, firmer coral limestone (Original)

tions them only once, in connection with minor coastal oscillations, when he asks, "whether perhaps those geologists may not be right who assume that the level of the ocean is subject to secular changes as a result of astronomical causes". But he did not take into consideration that because immense masses of water were held bound as continental ice, the level of the oceans during the individual glaciations was up to 200 m lower than it is today. During the post-glacial period of melting, it gradually rose again.

As the level of the water rose, corals kept pace by luxuriant growth, and thus produced the reefs of today. This glacial control theory was developed by Daly (1915), and although it is undoubtedly largely correct, it too was tightened up, modified and extended as more detailed knowledge became available on existing reefs and new types of reef.

Darwin himself acknowledged that where the sea is shallow for long stretches, irregular reefs are formed that cannot always be classified. This statement, as well as expressing a degree of uncertainty, also indicates the need for a continuing process of reef classification. For example, platform reefs that have grown on a submarine bank can become atoll-like structures, if, as the tide ebbs, the water accumulating on the back of the reef exerts a great weight, compressing the loose substratum formed by

the corals. The more stable margins where coral growth still continues are gradually forced upwards to make a ring-shaped embankment round a shallow lagoon. This explanation for the development of these "pseudo atolls" was called by Hass (see p. 13) the theory of central subsidence.

Darwin also had difficulties with the "submerged banks" of the West Indies which "cause considerable doubts as to their classification" (1899, p. 206). In 1970, Kühlmann developed the theory of obstructive and retarded growth for the reef area of the West Indies and the bank reefs that are widespread there. Their development is linked with particularly strong glacial influence. As a result of the binding of vast quantities of water as inland ice masses that extended across the North American subcontinent as far as Iowa, the water level during the various glacial epochs was 70 to 200 m lower than it is today, and annual average temperatures were 8 °C below today's values. So for a long time, reef-building corals were deprived of the basic conditions of existence. It was not until a thousand years after the end of the last glaciation that the oceans of Central America had warmed sufficiently for hermatypic corals, spreading in from the south, to be able once again to colonize the shores of the Greater and Lesser Antilles and suitable continental coasts.

The edges of the post-glacial abrasion plateau, that otherwise offered good conditions for colonization, now lay deep beneath the surface of the water that had risen as the ice caps melted, and the hermatypic corals, dependent upon light, could not grow there. So the corals settled in well-lit shallow water and initially built fringing reefs. After that, the corals along the outer edges of the reef kept pace by prolific upward growth with the continued rise in the level of the sea. At the same time, a shallow lagoon was formed between the land and the reef. In this way—greatly obstructed and retarded—the bank reefs lying in the middle of the coastal plateau were formed.

But bank reefs can also come into being passively by processes of abrasion. The limestone coasts of the West Indies often exhibit terrace-like characteristics that developed as a result of sea and land oscillations. If a submerged terrace is affected by turbulently moving water, its edge is immediately colonized by calcareous algae, corals and other limestone-depositing organisms that cover it with a living, highly resistant encrustation. But loose fragments and sand are continually cast upon the plateau and broken down by waves. The currents created by the constant pressure of new water masses across the plateau together with the action of gravel and sand, erode a depression in the surface of the rock that follows the direction of the terrace, and this depression is gradually enlarged. But the firm, resistant edge remains as a raised bank. Out of a coastal terrace, a bank reef has been carved, that is not a primary reef formed by corals but a secondary reef colonized by them.

As we have seen, coral reefs occur in a much greater variety of types than Darwin could have realized in his day. The diversity of individual reef types has developed in very different ways, depending upon the geology of the

particular reef. The basic idea of Darwin's subsidence theory—shallow, turbulent water provides the optimum conditions of growth for reef-building corals, and that they compensate for increased depth by constantly striving to grow up towards the surface—remains a significant one, even if details have had to be modified and supplemented.

To recapitulate and generalize: Since reef-building Scleractinia achieve a sufficiently high rate of growth and therefore production of limestone to be able to build reefs only in association with unicellular symbiotic algae of the species *Symbiodinium microadriaticum* that occur in large numbers in their living tissue, they are restricted to the photic zone of shallow water, because their symbionts, like every other chromatophoric plant, require light for the processes of assimilation. If the water covering them becomes deeper—whether as a result of subsidence of the ground, a rise in the level of the sea or a combination of both these factors—they are able, because of the optimal conditions of growth that normally prevail in reef regions, to compensate for the "height deficit" by means of vigorous growth. This is crucial to an understanding of coral reef formation. Since all theories on the development of coral reefs are based on these facts, they were brought together by Kühlmann in the compensation theory, published in honour of Darwin on the centenary of his death in 1982. The title implies both the geological process of gradually increasing depth of water covering the reefs and the biological process of simultaneous compensating upward growth by the corals.

Coral distribution in the oceans of the world

All living creatures make use of opportunities to conquer new biotopes, even ses-

81   50

50   *This Feather Star has delicate arms, rounded at the ends, which it holds extended even during the day, in dim light at moderate depths. (Red Sea)*

82/83   51 | 52

51   *The long flexible branches of this Horny coral* (Pseudoplexaura crucis) *curve like whips in the swell of the waves. (Virgin Islands, Western Atlantic)*

52   *Red and whitish Soft corals, with branching growth, of the genera* Dendronephthya *and* Lithophyton, *often colonize projections on steep rock walls. (Red Sea)*

84   53 54
     55

53   *The coarse, white spines of this Sea Urchin* (Echinothrix calamaris) *contrast sharply with the fine, dark-coloured subsidiary spines. (Red Sea)*

54   *As the Slatepen Urchin (Fam. Cidaridae) gets older, its thickened spines acquire an encrusting coat of algae and bryozoa, so it is well-adapted to its environment. (Red Sea)*

55   *The Starfish* Protoreaster lincki *is a voracious predator, eating everything slow enough to be caught by the suckers of its tube feet: molluscs, worms, crabs, other echinoderms and also coelenterates. (East Africa, Indian Ocean)*

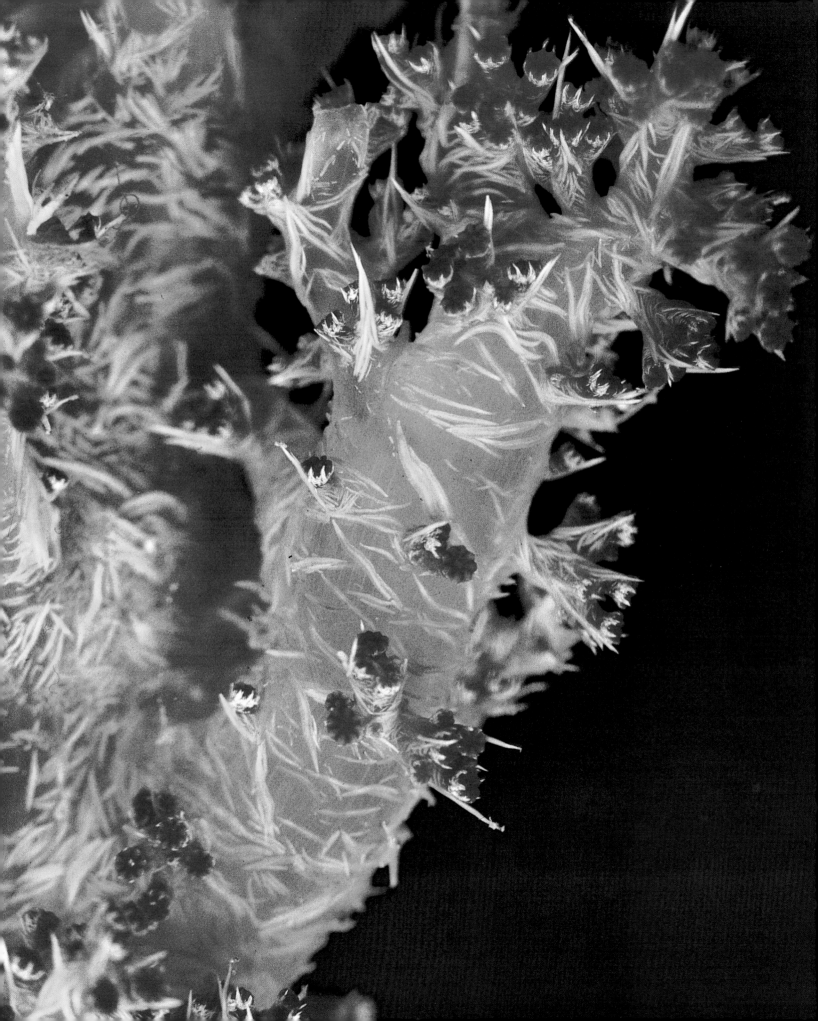

56 *The "Sea Mop"* (Thelenota ananas) *is a large Sea Cucumber that can be up to 1 m long. It occurs sporadically on areas of sand at considerable depths among reefs. (Society Islands, Pacific)*

57 *Large Thorny Oysters* (Spondylus varius) *are found primarily on steeply sloping reef walls in the calm area of the lagoon. The thickly-encrusted valves are open, and the animal's brightly-coloured mantle is clearly visible. (Society Islands, Pacific)*

58 *In the nutrient-rich waters of lagoons, small Tridacnas* (Tridacna crocea) *settle in such large numbers that the inhabitants use them for fish bait and as food. (Tuamotu Archipelago, Pacific)*

59 *The small, nocturnal Long-tailed crab* (Enoplometopus occidentalis) *inhabits hollows in the porous limestone of the reef. (Hawaii, Pacific)*

60 *Spirobranchus giganteus, a "Feather-duster" Polychaete worm, occurs in a wide variety of colours—white, yellow, red, blue or dappled. (Red Sea)*

61 *Soft corals of the genus* Dendronephthya *usually have a fairly thorny texture because of the large, hard, calcareous sclerites that project from them. (Red Sea)*

62 *Nudibranchs are snails without shells; like this* Chromodoris quadricolor, *they often exhibit brilliant markings—perhaps a warning coloration. (Red Sea)*

sile ones—sedentary animals in water, plants on land. They spread by means of spores and seeds that are borne on the wind or carried by animals. The floating larvae of sponges, bryozoa, tube-dwelling worms and corals are carried along in water. Where did the stony corals come from and how far were they able to spread? The answer is provided by the great currents in the oceans of the world.

In the early Tertiary, when the mighty Tethys covered the world's tropical belt, the coral species were fairly evenly distributed. It was not until the mid and late Tertiary that they were split into the Atlantic and Indo-Pacific regions by the land bridge linking North and South America and the area of land that linked North Africa and Asia.

From that time onwards, two different faunal zones developed, of which the Atlantic was affected later by glaciation, while the Indo-Pacific largely escaped it. Many Tertiary species of coral are still widely distributed in the Indo-Pacific, whereas the Atlantic representatives have died out. In the Indo-Pacific region, there are more than 200 species of *Acropora*, but in the Caribbean only three, while more than 30 Indo-Pacific species of *Porites* contrast with only six Caribbean. There are two principal causes of the extreme uniformity in the bio-geographical character of the vast Indo-Pacific coral reef region that extends from east to west across some 25,000 km. Firstly, the stability of the biotope, in which the environmental conditions have remained essentially unchanged over a very long period of time, made new mechanisms of adaptation unnecessary. The essential stimulus for the creation of new species did not exist. Secondly, the larvae of stony corals are capable of living in a planktonic state for some weeks, in certain cases up to two months. Consequently they can drift in ocean currents for thousands of miles before they settle and develop colonies. As a result, colonization was

fairly uniform across the whole of this extensive region (Fig. 25).

The post-glacial recolonization of the cold Atlantic region by hermatypic Scleractinia was brought about by larvae drifting northwards only after the water masses, starting in the south, gradually became warmer. But the area in which stony corals could settle was interrupted off the coast of South America for a distance of about 3,000 km by the Amazon and Orinoco Rivers that poured their cloudy, sediment-laden waters into the Atlantic. To the south, the independent Brazilian coral zone came into being, with a reduced coral fauna of predominantly endemic species. Here, different environmental conditions existed, leading to the development of genetically modified new species. On the other hand, the Bermuda Archipelago, situated a good deal farther north, includes only species that also occur in the Caribbean, the larvae of which have been carried there by the Gulf Stream. Thus it does not represent an independent coral zone but rather an impoverished one.

In the littoral of the West Indies today, 90 per cent of the Horny corals, 80 per cent of the round crabs and so some 300 species, 80 per cent of the Sea Cucumbers and all the Crinoids are endemic. Of the Parrotfishes, the genus *Sparisoma*, with 18 species, is endemic. The great wealth of endemic species adapted first to altered and then to optimal conditions.

On the other hand, the separation of the Atlantic from the Indo-Pacific zone is of too recent date to have obliterated all relational links in the fauna. For example, of 26 reef corals native to the West Indies, 21 genera also occur in the Indo-Pacific. But the species are not identical. The absence of the many species of the genus *Acropora* in the fringes of the two reef regions—in the eastern Pacific, off Hawaii, the Marquesas Islands, western Australia, the Bermudas and the coast of Brazil—is a shared feature that must be seen as the result of ecological factors. As a true evolutionary product of the tropical littoral, all species of Staghorn coral are so specialized in their temperature requirements that the peripheral areas are simply too cold for them. The 500 or so species of hermatypic stony corals found in the Indo-Pacific are an indication of the enormous potentialities for development that this part of the ocean, twenty times the size of the Atlantic and scarcely at all affected by glaciation, has been able to provide for corals. In both regions, the principal currents flowing from east to west, and carrying with them planktonic larvae, have led to a maximal concentration of species in the western part of the Pacific, the Indian Ocean and the Atlantic. Drifting fragments of pumice on which small coral colonies have settled have undoubtedly also contributed to the distribution of corals. Endemic species have also developed in the enclosed biotope of the Red Sea, where increased salinity and reduced hydrodynamic activity have produced unique ecological conditions.

Of course, faunistic differences have also affected recent reefs to a considerable extent. For example, Atlantic reefs exhibit smaller forms than Indo-Pacific reefs, and are less capable of withstanding the buffetting of the waves—the first because of the retarding effect of glaciation on growth, the second because of the smaller total of reef-constructing organisms and particularly the reduced number of species of "cementing" calcareous algae in the surf zone. Linked to this, the percentage area covered by corals also varies. While in the Caribbean it is about 60 per cent, in the majority of the Indo-Pacific reefs it amounts to 80 to 90 per cent. Since in many places the corals are extremely densely packed together and the species grow one above the other in strata,

according to their sizes, it is tempting to assess such a degree of cover as more than 100 per cent.

Not all reef substrates are utilized to an equal extent by stony corals. In the Indo-Pacific reef region, together with representatives of the genera *Psammocora, Montipora, Favia, Goniastrea, Prionastrea, Leptastrea, Echinopora, Podabacia, Cyphastrea, Porites, Pachyseris, Pavona, Leptoseris, Coscinarea, Hydnophora* and others, many species with encrusting or disc-shaped growth are able to colonize even quite steep walls. In the Atlantic region, only a few species of the genera *Madracis, Siderastrea, Agaricia* and *Leptoseris* are of this kind.

Differences in colonization can be seen elsewhere, such as in the coral growth on reef crests. In the Caribbean region, Elkhorn coral, *Acropora palmata,* is dominant, accompanied by Fire coral, *Millepora complanata.* No type corresponding to the Elkhorn coral occurs in the Indo-Pacific region. Nor is *Millepora* usually found in such quantities as in the West Atlantic. Instead, bush-like Staghorn corals grow in the Indo-Pacific in the constantly churning waters of the reef crest; they are convergent species of the Caribbean *Acropora prolifera,* although the latter occurs exclusively in calm water. In places, associations of *Montipora* of uniform specific composition extend across the outer slope and the reef ridge at a depth of 2–20 m. The edges of their fragile plates and dishes gleam brightly in the gloomy depths, giving them a particularly striking appearance. It is always surprising to see these extremely fragile *Acropora* and *Montipora* corals growing in turbulent waters without suffering any damage.

Differences also exist in rhythms of activity (see p. 106). While in the West Indian province, only *Dendrogyra cylindrus* is obligatorily diurnal, *Montastrea* and *Porites* occasionally diurnal, in the

Indo-Pacific there is a whole series of diurnal corals such as *Porites, Galaxea, Goniopora, Favia* and others. Whether this is because the competitive struggle for food is more intense there, remains an open question. Perhaps nocturnal activity is not sufficient to supply the amount they require.

There are striking differences in the incidence of Octocorallia. In the Indo-Pacific reefs—apart from those situated in the central and eastern Pacific—well-developed colonies of Soft or Leathery corals occur frequently in large numbers. In the Atlantic reefs, Alcyonacea are not entirely absent, but are found only sporadically and sparsely. Here instead, innumerable large, brown, violet and yellow bushes of Gorgonians sway gracefully in the ocean swell. In the Indo-Pacific, at least in the shallow water zones, they are much reduced.

In the tropical West Atlantic, the number of species within particular groups of animals is lower than in the Indo-Pacific. For example, in the Indo-Pacific reefs there are some 5,000 different gastropods and bivalves, while in the Atlantic, perhaps 1,200. About 2,000 species of fish compare with some 600. 140 species of Butterflyfishes (Chaetodontidae) live in the Indo-Pacific, only 12 in the Atlantic; the Indo-Pacific has 65 species of Angelfishes (Pomacanthidae), the Atlantic 9. In the oceans of Central America, the genera *Dascyllus, Amphiprion* and *Premnas* are not represented, and Eibl-Eibesfeldt is undoubtedly correct when he suggests that cooling which occurred during the glacial epochs drastically reduced the fish stocks there. The more closely linked the fish were to the coral reefs, the more certain were they also to be killed by the cold. But any falls in temperature in the Indo-Pacific during the periods of glaciation were so minimal that they had no serious effect on the organisms there.

Motley life

in the living reef

Coral reefs represent the only biotope in the world that is created, enlarged and renewed by the community of organisms particular to it, which themselves supply energy and materials. In other words, the coral biocoenosis is largely autochtonous because it itself creates and maintains its biotope, the coral reef. In doing so, it provides the basis of existence for a wealth of other organisms. In the course of evolution, all of them have adapted to one another so successfully that normally no one species is eliminated by another. They have achieved a state of biological balance.

It is not possible to describe in detail this overwhelmingly rich system of individual forms of life, for two reasons: firstly, we still know too little about most of the inhabitants of the coral reef to be able to generalize, and secondly, even what we know would prove too voluminous to present in detail here. Perhaps some selected examples may serve to illustrate vividly the principal characteristics.

It should, however, be noted that although the coral reef differs from other ecological systems as a result of its capacity for autochtonous reproduction, it is yet entirely incapable of existing isolated from the universal environmental factors of our world, sun that gives light and energy, air that provides oxygen, and the life-spring, water, with its vital solutes. If any one of these factors is altered, even slightly, the ecological structure of the coral reef collapses.

Biotope coral reef

Coral reefs come into existence where hermatypic stony corals in tropical oceans find conditions favourable to them: rocky ground in clear, turbulent water. In coastal reefs, windward and leeward sides are quite distinct. Oceanic coral reefs also exhibit these features but in a different way: an atoll has an area of calm water in its lagoon.

According to the extent of its exposure, a reef can be divided into the outer reef that is open to the sea and the protected inner or backward reef. The outer reef consists of the reef ridge that is usually slightly heightened by the growth of calcareous algae and corals, the adjacent outer edge, the weatherward or outer slope that extends down into the sea, and the fore reef. If the outer slope extends a considerable distance down into calm water, it can be divided into a wave zone, an eddy zone and a current zone. The inner reef, on the other hand, has only a broken reef edge and a small leeward slope, also called the inner or rearward slope. The horizontal area, that can vary in width from a few metres to hundreds, lying between the outer and rearward reef, is called the reef roof, plateau or flat. The edges and the reef ridge lie at the mean water level, the reef flat at the low water level.

These sections of reef are variously structured. The seaward slopes that are affected by heavy seas are often characterized by grooves perpendicular to the shore that have been cut into the limestone rock by the action of gravel and sand. Together with the ridges that remain between the grooves and extend like spurs or tongues out into the sea, they are known as a groove and spur system. If the reef flat is so narrow that surf breaks across it violently as it flows shorewards, a groove and spur system can also develop there. Moreover, the action of rushing water and sand can grind out channels several metres deep in the foot of a reef that lies close to the surface. Caves have been washed into the rock wall, particularly at points that formerly lay at sea level. Everywhere, the porous limestone rock is cut through by crevices and fissures. The entrances to caves that can be quite small or as tall as a man, frequently front the open sea a few metres below the waterline, extending through the rock to emerge on the reef flat. Inflowing waves force water violently through these blow holes, sending great fountains spraying up into the air. Pebbles collecting on ledges and terraces grind at the rocks like mills. Projections, niches, domes and sills produce every conceivable conformation of rock surface, that is colonized and inhabited by organisms in a great diversity of ways. Depending upon the extent of its exposure to water movement, it is taken over by organisms that enjoy turbulence or those that prefer a quiet environment, and depending upon conditions of light, by species that prefer the light or the dark. A multiform surface is always occupied by an association of varied organisms. But if it is uniform, a single species may predominate. Other prevailing environmental factors, such as heavy surf or sedimentation, can result in uniformity of colonization, since relatively few organisms are capable of adapting to extreme conditions.

What is life like in the reef?

Konrad Lorenz, a Nobel Prize winner, stated that in respect of the diversity of its living creatures, no biotope on earth can surpass the coral reef, and no other poses such a diversity of ecological problems (1976). But how has it come about that representatives of every phylum in the animal kingdom and of almost all orders of plants, from the amazingly rich sphere of micro-organisms, invisible to the diver, to large forms that occur as solitary as well as colonial animals, have been able to exist for millions of years in many thousands of different species within such a small space, without this ecosystem faltering in any way? The coral-zooxanthellae symbiosis, as a biological force, is the most important factor in its regeneration (see p. 44), in its longevity and as the basis of existence of myriads of reef inhabitants. But this symbiosis would have remained ineffectual, had not the organisms harmonized their behaviour in such a way that one species tolerates the presence of the other. Their behaviour is simultaneously an adaptation to the inanimate and to the animate environment. Adaptation also demands that no species shall make precisely the same bio-demands as another, otherwise the weaker would have to give way to the dominant species.

The coral reef in its three-dimensional expanse, with horizontal, diagonal, vertical to overhanging substrates, level stretches, gorges, terraces, caves and hol-lows, pedestals and clefts, with sunshine, shade and darkness, with windward and leeward aspects, with the progressive decline in light and water movement from the surface downwards, offers the marine organisms a uniquely varied supply of habitats (Ill. 32). A host of sessile organisms has accepted the challenge and settled there.

The luxuriant growth of coral is accompanied by vast numbers of calcareous algae, bryozoa colonies and hydrocorallids. Here, as in a forest, growth occurs in tiers; the taller organisms rise above the shorter ones. The growth layers or biostrata of the coral reef correspond to those of a forest on a scale of more or less 1:10 (Fig. 27).

1. The soil horizon of the forest is inhabited primarily by vermiforms, larvae and insects. In the reef, mushrooms, algae, sponges, bivalves and annelids bore into the layer of rock that is the substratum.
2. In the forest, this is followed by the moss layer. In the reef, there is an encrusting layer about 1 cm thick, formed by calcareous algae, sponges, bryozoa and stony corals.
3. The forest's herbaceous stratum corresponds to the next layer in the reef—that of the small corals. This is made up of a multiplicity of different Scleractina, together with delicately branching bryozoa, fragile hydrocorals and tufted algae, growing to a height of 1–10 cm.
4. The shrub stratum of the forest compares with a similar shrub stratum in the reef. It is composed of quite large, branching corals, predominantly Staghorn, Styloid and Claviform corals. As in the forest, they create a dense "impenetrable undergrowth".
5. Trunks of trees grow up out of the herbaceous layer, and so too do those of large corals, to form the trunk stratum. And like the wood pests in tree trunks, sponges, bivalves and annelids bore into the stems of corals.
6. Over it all stretches the canopy stratum, consisting in the forest of treetops, in the reef of coral heads—the large umbrellas of Staghorn coral, or the upper, living parts of huge pillars of Pore coral.

Neither in woods nor in coral reefs do the strata usually occur as completely as described here. But in both cases, however coincidental the similarity may be, this kind of stratified growth achieves the same end: the full utilization of the area colonized, and a large number of ecological niches and habitats for many species and individuals. The sessile organisms have divided between them the varied surface of the reef, according to their needs, in the most efficient way possible. Because of their growth forms, they increase the diversity of the substratum, and offer extra living space to other creatures. And if, tentatively, one were to compare the crustaceans of the reef to the beetles on land, then the fishes might appear rather like the birds. Indeed, they are the most striking inhabitants of the reef.

Lively competition for space greatly restricts attempts to expand on the part of those organisms incapable of locomotion. The colony must remain permanently in the place on which it settled initially. If as a result of erosion and the death of a colony, a site suitable for colonization becomes free, and no coral planula is at hand and ready to settle, it is normally taken over by algae. Repossession of the site may take years or tens of years, although it is usually achieved in the end by specialist organisms such as the Scleractinia, so that finally the "opportunists" are displaced by the "specialists". But when the biotope is changed permanently as a result of human intervention, recolonization is impossible.

On the other hand, suffocating organisms such as algae, sponges or soft corals are often able to cut off the supply of light and water from corals and grow

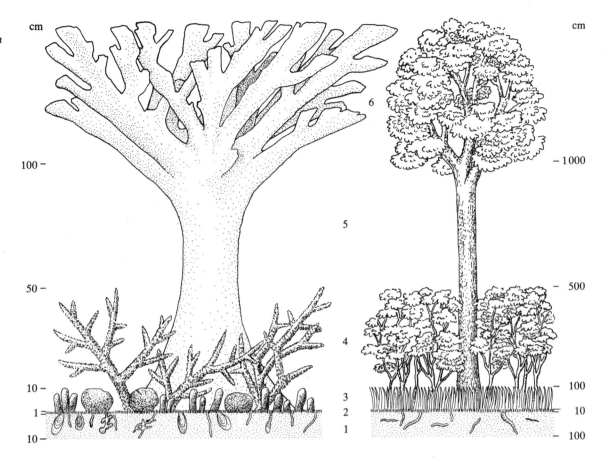

27 Coral reef and rain forest make
optimal use of the biotope, creating strata
with a great many ecological niches.
1 Limestone rock-/soil stratum,
2 encrusting-/moss stratum,
3 small corals-/herbaceous stratum,
4 bush stratum,
5 stem or trunk stratum,
6 canopy stratum (Original)

over them (Ill. 43). Many different factors can cause this to happen. Fishes that
feed on algae are unable to enter certain
territories that are defended actively by
reef perch. As a result, the algae multiply
to such an extent that they suffocate the
stony corals. A few years ago, so much
municipal effluent containing nitrogen
and phosphorus was discharged into the
Bay of Kaneohe in Hawaii, that there
was a population explosion of algae
which destroyed the corals. Off Guam, a
sponge of the genus *Terpios* spread its
encrusting growth over colonies of stony
corals that had previously been predated
by Crown-of-thorn starfishes (see p. 125).
In coral reefs that have been subjected to
heavy sedimentation as a result of construction work or blasting operations,
Soft corals that are tolerant of silt, frequently develop in large quantities. They
pump themselves up by taking in water
and suffocate the stony corals.

Stony corals also contend among themselves for possession of the substrate.
Some of them are furnished with long,
extensible polyps, others with whip-like
tentacles, many with long mesenterial filaments protrusible from the gastro-vas-
cular cavity, all are armed with batteries
of stinging cells. In this way, they maintain a clear space round the colony and
repulse weaker Scleractinia. Since the
mesenterial filaments secrete corrosive
ferments, competitors coming into contact with them are partially digested and
die. The species with the longer filaments
usually vanquishes that with shorter
ones. A form of "pecking order"
develops among the stony corals (Schuhmacher, 1976). The advantage is on the
side of those species that require only a
small base on which to grow, but which
then spread well across it. This is certainly one of the reasons why branching
corals often have extensive, vigorous pop-

ulations and table-like, widely-extending
colonies.

The wide range in the quality of conditions and the large number of competing
species have compelled reef organisms
constantly to develop new forms of specialization in order to exist together. So
species have constantly occupied new
ecological niches and found fresh ways
of securing an existence, resulting in the
enormous diversity that the coral reef
exhibits today. The term "ecological
niche" is not restricted here to its spatial
aspect, but also includes other functions
vital to existence—locomotion, feeding,
reproduction, defence—that are coordinated in successful adaptive behaviour.
An additional feature is that, in spite of
this synchronized interaction, an increasing number of species creates stronger
competitive pressure. Many species solve
this problem by splitting into new species with altered behavioural patterns.

Once again they are able to exist success-fully, but at the same time they increase the competitive pressure throughout the whole community, providing others with the incentive for renewed radiation. The coral reef is, as it were, a catalyst in the development of a multiplicity of species, with both biological and physico-chem-ical factors functioning as control mechanisms. It is a process virtually impossible to measure, and which extends over millions of years. Hennig's statement that the number of species within a group of organisms is propor-tional to the age of that group has its analogy in the sphere of ecology: the number of species in an old biocoenosis is higher than in a young one—coral reefs are among the oldest ecosystems in the world.

While the sessile organisms live on the solid substrate, the vagile organisms inhabit both the ground and the water above it. They can be divided into two groups, the motile benthic organisms that live on the ground, in short called vagi-benthics (L. *vagus,* wandering, Gr. *ben-thos,* depth), and the plankters or plank-tonts that live suspended freely in water, known collectively as plankton (Gr. *planktos,* drifting). In between these two groups are many species that live sometimes on the ground, sometimes suspended in water.

Many vagibenthics are confined to living on or in the ground. The solid sub-strate is inhabited by boring organisms: blue-green algae, sponges, annelids, sipunculids, molluscs and echinoderms. Many boring animals have a degree of independent movement within the cavi-ties or channels they inhabit. Blue-green algae spread a fine, often dense network to a depth of several centimetres within limestone. It has been estimated that, locally, these minute algae dissolve more than 80 per cent of the limestone with their acid-like secretions. Boring fungoid mycelia have also been found in the rock of coral reefs.

Below a depth of 20 m, the light is so weak that the numbers of boring algae diminish. They are replaced by clionids or boring sponges (Clionidae). These have corrosive cells that extend processes onto the rock, deflate and release a caustic substance. The cells are immedi-ately replaced by new ones that continue the corrosive process, producing a system of cracks up to 0.2^{-6} m wide, and finally a tiny particle of loosened limestone 30–50^{-6} m in diameter. This is carried away in a current of water set up by the sponge. *Cliona lampa* is able to destroy some 6–7 kg limestone in a square metre of reef surface in 100 days (Neumann, 1966). Since only about 2–3 per cent of the material bored is dissolved, a consid-erable proportion of the debris deposited at the foot of the reef results from the activity of boring sponges. In the reefs off Jamaica, they are the major agent of erosion at depths below 25 m. Sipunculid worms appear to bore their way into limestone rock in a similar manner. They also loosen crystals of limestone piece by piece and move them aside mechanically. But the problem of how some annelids construct their U-shaped tubes in lime-stone rock still remains to be solved.

In the limestone of the reef, there are large numbers of rough holes constructed and inhabited by bivalves. The common Sea Date *(Lithophaga)* bores by means of a chemical process. Glands on the siphon and edges of the mantle secrete strong acids by means of which the crea-tures bore into limestone. The diameter of the hole is approximately that of the longish-oval shell, the front of which faces the opening. Out of it, the siphon is extended, a tube-like structure through which cilia activate a flow of fresh water carrying plankton and oxygen into the mantle cavity to supply food and permit respiration. Rock-boring clams (Phola-didae) prepare their homes in a different way. The lower, thicker end of the shell is armed with teeth. The muscular foot rotates it constantly in one direction.

Although movement is very slow, never-theless over a period of time and as the bivalve grows, a hole is formed. Other rock boring bivalves include a species of Giant Clam, *Tridacna crocea,* that bores in the Pacific reefs.

In the surf zone of the reefs, there are some rocks so full of large round holes that they resemble Swiss cheeses. Hump-backed Sea Urchins are the creatures responsible, and some of their spines may be seen protruding into the open water. The pockets in which they live fit snugly round them. The most violent breakers cannot dislodge them as they adhere firmly with their sucker feet and wedge themselves into the rock with their spines. Particles of organic food are washed in to them. The spines protect them from enemies. It is not known for certain how the holes are constructed, but since they also occur in sandstone, gneiss, lava, slate and granite, it can be assumed that the very hard teeth and spines abrade the rock mechanically. Boring genera of sea urchins are *Echino-metra, Echinus, Arbacia, Cidaris* and *Heterocentrotus.* The latter are furnished with powerful three- or four-edged spines, the tips of which have curved and serrated cutting edges, making them for-midable tools for grating and boring. The largest spines on the urchin's cal-careous test are situated where its diameter is greatest. So the pocket in which it lives is at its widest halfway up, while the opening at the top is narrower. Consequently, the sea urchin can never leave its home again. Of two sympatric Hawaiian *Echinometra* species, *E. oblonga* lives in turbulent water, *E. mathai* in calm. Competition that could endanger survival is thus avoided (Ill. 54, 89).

Taken as a whole, reef-boring orga-nisms represent an effective component in the formation of the rock mass. In the reefs round Low Island in the Great Bar-rier Reef, drilling organisms were found to include twelve species of bivalve, 1 of crustacean, 2 of sponge, 2 of sea urchin, in addition to fungi, algae, sipunculid and annelid worms (Otter, 1937). In limestone of the Bermuda reefs, 24 endolithic animals were noted: 8 sponges, 7 bivalves, 3 annelids, 2 sipunculids, 2 balanids, 1 snail and 1 sea urchin (Bromley, 1978). Although a considerable proportion of the substance of the reef is destroyed by boring, the loss is more than offset by the growth of organisms laying down limestone skeletons, in particular of stony corals.

In contrast to the drilling organisms that are strictly linked to a site, thousands of creatures move freely through the holes and caverns of the reefs. Underwater caves always house a collection of creatures that shun the light. Here, the sessile organisms exhibit distinct zoning. Close to the entrance are grouped sponges, hydrozoa, gorgonians, ascidians and algae, all of which prefer a certain degree of light and moving water. In the dim recesses of the interior, however, are those species that prefer still water and deep shade. A few vagibenthics live on the walls of the caves. They include predatory ribbon worms (Nemertini), nudibranchs or naked-gilled sea slugs and Firefishes (Ill. 135, 137). But organisms that move freely in open water are also common in caves, including swarms of minute, transparent Mysid Shrimps, small Copper Sweepers (Pemphris) and Cardinalfishes (Apogonidae). Caves are particularly favoured by fishes that require rest. Since rest phases recur at regular daily intervals, fishes occupy the caves only at particular times, for example, Sweetlips (Gaterin gaterinus), Soldierfishes (Priacanthidae) and Rock cods (Serranidae) in the middle of the day, and at night Butterflyfishes and related species (Chaetodontidae, Pomacanthidae). Here many of the fishes exhibit a characteristic light-orientation reflex. Since in swimming, they normally keep the back directed towards the light and the belly towards the darkness of the

ground or the depths of the sea, when they are near the roof of a cave, they swim with the belly turned upwards, because the roof is darker than the interior of the cave that is illumined from the entrance.

Many creatures, solitary in habit, avoid the larger caves but prefer the safety of small cavities, passages and crevices that are common in the porous rock of the reefs (Ill. 59). They are mainly fishes, crustaceans, molluscs, vermiforms and a few actinians. Crayfish extend their powerful, spine-bearing feelers, Moray eels their dagger-like teeth from the holes in which they live. If disturbed, Gobies (Gobiidae) and Blennies (Blenniidae) disappear completely.

Many species take up residence in empty snail shells or mussel shells, and the tiny Blenny, Plagiotremus townsendi, even in the casings of Worm-shells and Tube-dwelling worms. Hermit crabs also conceal their soft abdomen in empty snail shells (Ill. 44, 140). When this house becomes too cramped for them, they are forced to seek out a larger one. But since empty snail shells are not always at hand, some rivalry may result. It is sometimes possible to watch land-dwelling Hermit crabs (Coenobita), that often live in large numbers on the beaches behind reefs, engaged in fighting for the possession of a larger shell. They climb nimbly among the pebbles on the shore, sometimes even in bushes (Ill. 79). Among the creatures that inhabit coral islands, mention must be made of the Robber or Coconut crab (Birgus latro) (Ill. 77, 78). It is some 30 cm long. It makes use of a snail shell as its residence only when it is quite young. The abdomen of the adult crab is protected by rigid bands. It lives in caves in coral rock or under trees and prefers fatty food such as dead animal matter, coconuts and pandanus nuts. Because it is considered a delicacy, it is hunted relentlessly, so that in many places, it has already become extinct.

Certain species of fish, such as Jawfishes (Opistognathidae) and the ethereally delicate, pastel-coloured Sleepers (Eleotridae), construct tubes in the sand in which to live. The Jawfishes line them skilfully with pieces of coral debris fitted one against the other. Striped Catfishes (Plotosus anguillaris) live closely confined under massive coral colonies in cavities that they themselves have dug out of the sand. As more sand trickles in, they carry it away in their mouths. They are very timid and are rarely seen by divers.

Some organisms, such as actinians and octopuses have very plastic bodies. When they are looking for a hiding place, they are able to make their body extremely thin, so that it "flows" into narrow gaps and cracks. Whereas the actinian rarely emerges again, the octopus may come out at night and rove about in the vicinity. Brittlestars live under boulders, while small Cones (Conus), Cowries (Cypraea) and Rock shells (Murex) are found among pebbles. Great Tritons or Sea Trumpets (Charonia), Fighting conches (Strombus), Spider shells (Lambis), Tops (Trochus) and Helmets (Cassis) crawl across the rocks (Ill. 100). Their shells, like those of quite large bivalves that are attached to rocks, such as Thorny Oysters (Spondylus), Noah's Ark mussels (Arca), Pearl Oysters (Pinctada) and Giant Clams are covered over by a dense growth of algae (Ill. 57, 58). Strict territoriality is shown not only by benthic fishes and crustaceans, but also by Cones and many other "lower" organisms. On the outer slopes of the Red Sea, crinoids exhibit a distinct division into zones in that only particular species occur in shallow water, in moderately deep and in deep water.

Adhering to rocks in the surf zone are oval-shaped, slightly hump-backed organisms with protective plates, the chitons or loricats (Placophora). Even the most violent breakers are unable to tear them away, particularly since their adhesion is supported by secretions from glands in the broad sucker-foot. But if they should occasionally lose hold, they roll themselves up into a ball like a wood-louse, and are tossed harmlessly across the rocks by the waves. Large numbers of limpets (Patellacea), which fit the edges of their shell tightly to the base on which they rest, and stalkless balanids attach themselves to rocks. Many creatures show a distinct migratory instinct. Spiny lobsters assemble in large numbers in the Caribbean—much as migratory birds do—at particular seasons of the year, before setting out on migratory journeys of many miles. Many Starfishes and Sea Cucumbers (Cucumaria) migrate, while echinoids tend to remain in one place.

Quite often, broad sandy flats extend from the foot of the reef or at the edge of the inner reef. In the zone of surface waves, the limestone sand is constantly stirred up and turned over, as ripple marks indicate. Few animals live in this extremely abrasive environment (Fig. 32). The areas of sand appear desolate and deserted. In calm water, the picture changes. Small hills and valleys announce the presence of annelids, specialized Sea Cucumbers and Pogonophorans that live beneath them in U-shaped tubes. Particles of food slide down into the funnel-shaped valleys and are eaten by the animals. From the small hill that covers the creature's hind end, tiny clouds of faecal matter are ejected from time to time, like a miniature volcanic eruption. Tube eels (Heterocongridae) as thick as pencils, live in tubes out of which only the front end of the creature projects. They are the only fish to have become almost completely sessile. Crabs often construct branching galleries that are more complicated than a fox's earth. Others camouflage themselves by burrowing into sand. Flatfishes and Rays also behave in this way, covering their body with sand by deft movements of the fins. Large numbers of gastropods and

bivalves also live in sand. Quite large Sea Pincushions and Burrowing Sea Stars and multitudes of holothurians (Holothuria) lie exposed on the surface of the sand. Large snails crawl over the ground. Sand-coloured Red Mullets (Mullidae) move gently above them.

A number of large green algae, lobular or brush-like in form, extend fragile, root-like organs into loose sand, cement grains of sand to these rhizoids and in this way, obtain a firm hold. Sea grasses form extensive underwater meadows. They all moderate the movement of the water and protect loose ground from constant shifting. In this way, these plants make it possible for many more organisms to exist. Moreover, large numbers of microscopically small organisms settle on the algae and sea grasses, providing young fishes in particular with food and a place of refuge.

All the fishes living in the coral reef depend in some measure upon the substratum (Fig. 28). Slow-moving, sea-bottom fishes such as Flatheads *(Platycephalus),* Stonefishes *(Synanceja),* Scorpionfishes *(Scorpaena)* and the curious Batfishes *(Ongcocephalus)* are characterized by their unwieldy shape. Very lively by contrast are the short, robust, thickset Damsel- or Demoiselle fishes (Pomacentridae) and the small, elongate wrasses (Labridae), of which several species occur in every reef in shoals of thousands. Butterflyfishes (Chaetodontidae) and Angelfishes (Pomacanthidae) are narrow and high-backed. They remain close to the substrate, and if danger threatens, disappear instantly into holes and fissures. Sea bass remain all day almost motionless in sheltered places (Ill. 129). It is not unusual for sharks to rest in hiding places on the sea bed. Large Parrotfishes (Scaridae), the gigantic Humphead wrasse *(Cheilinus),* Unicorn fishes (Naso), Barracudas (Sphryraenidae) and swarms of Surgeonfishes (Acanthuridae), also known as Doctorfishes or Tangs, move freely

97 63

63 *Ring-shaped islands can also develop as a result of stagnation, the death of corals and sinking in the inner parts of large patch reefs, where only the outer edges continue to grow, as these faros illustrate. (Maldive Islands, Indian Ocean)*

98/99 64 | 66
        65 | 67

64 *Fringing reefs extend directly from the shore and out into the water. (Great Barrier Reef, Australia)*

65 *Bank reefs situated in the middle of the coastal plateau are widespread in the Central American Archipelago. (Virgin Islands, Western Atlantic)*

66 *A true atoll, of which a section is shown here, arises as a result of the subsidence of a rocky island, surrounded by a coral reef. Quantities of sand and debris are deposited on the reef that lies just below the surface of the sea, eventually forming islands that support green vegetation. (Tuamotu Archipelago, Pacific)*

67 *Huge barrier reefs a considerable distance offshore extend for hundreds of kilometres along the eastern coast of Australia. (Great Barrier Reef, Australia)*

100 68 |
        69 |

68 *The small island of Maiao, like many other Polynesian islands, is surrounded by a rampart reef. If an island sinks below sea level, an atoll is formed. Every day, large thermal clouds develop over the islands in the ocean. (Society Islands, Pacific)*

69 *Low Island is a small island that has developed on a large platform reef. (Great Barrier Reef, Australia)*

70 *Elevated coral limestone often forms terraces along the ocean shores. (Society Islands, Pacific)*

71 *Massive occurrences of tube-dwelling worms can lead to minor reef development even in northern seas. Here, Ficopomatus enigmaticus has built a thick ring of dwelling tubes round the base of a pale. (Mouth of a river on the Bulgarian coast of the Black Sea)*

72 *Stromatolite reefs were widespread in the ancient oceans that existed on earth many millions of years ago. Today, only occasional small live ones are found, of which the Stromatolite reef in the hypersaline Shark Bay of western Australia is the most impressive.*

73 *Mangrove bushes (Rhizophora sp.) put out prop or brace roots as an adaptation to variations in water level. (Society Islands, Pacific)*

74 *Plants prolific on saline ground have the juicy, fleshy leaves of the succulent as a means of water storage, like this Sesuvium portulacastrum that is found worldwide on tropical coasts. (Virgin Islands, Western Atlantic)*

75 *In contrast to the majority of seabirds, the White-capped Noddy (Anous minutus) nests in trees. (Great Barrier Reef, Australia)*

76 *Here, Silver Gulls (Larus novaehollandiae) are resting in the midday sun on rocks on the shore. (Great Barrier Reef, Australia)*

77 *A tiny larval robber crab in the glaucothoic stage uses a snail shell to protect its soft abdomen. Like this, it clambers ashore. (Marshall Islands, Pacific)*

78 *The mighty Robber Crab (Birgus latro) is one of the land-dwelling hermit crabs; it feeds on the fatty flesh of coconuts. (Marshall Islands, Pacific)*

79 *The Land Hermit crab (Coenobita perlatus) returns to shallow water only to deposit its eggs. (Hawaii, Pacific)*

80 *Land crabs often live many kilometres from the sea, returning there only for reproduction when they make their way in hordes into the ocean. (Cuba, Caribbean Sea)*

81 *A view from the steeply-sloping shore of the volcanic island of Moorea across the lagoon to the rampart reef on which the ocean swell breaks. (Society Islands, Pacific)*

through the water. The blue and red striped Fusilierfishes *(Caesio)* venture out into the sea off the outer slope. But in the evening, all return to the reef.

The rule that the more diversified the substrate, the greater the wealth of species and individuals also applies to fishes. In an East African reef region, using equal quantities of dynamite, 3 fishes were caught on a comparatively bare reef platform, while on the outer slope with its many caves, projections, clefts and a luxuriant growth of coral, the total was 101. Fishes with similar habits have divided up the coral reef biotope between them, as in the case of Damselfishes in the Red Sea. There, *Amphiprion bicinctus, Dascyllus aruanus, D. marginatus* and *Pomacentrus trichurus* live in the lagoon canal that lies towards the land. The inner slope is inhabited by *Pomacentrus sulfureus, Abudefduf melanopus,* and *A. melas.* On the reef platform there are *Pomacentrus albicaudatus, P. tripunctatus, Abudefduf lacrymatus, A. leucosoma* and *A. annulatus.* The upper part of the outer slope has been taken over particularly by *Abudefduf saxatilis, A. sexfasciatus* (Ill. 86) and *A. leucogaster,* the lower part by *Abudefduf azysron, Chromis dimidiatus, C. caeruleus, C. ternatensis* and occasionally *Dascyllus trimaculatus.*

Even the most diversified of substrates would not provide sufficient living space for the extremely abundant fauna of the coral reef, had not its members developed a daily rhythm in the form of a "shift system", as an adaptation to the "housing shortage". It entails the apportionment of phases of activity and rest within a 24 hour cycle (Ill. 59). At night, increased numbers of plankters rise to the surface from very deep waters. It is at this time that the many plankton-consuming animals leave their hiding places where they have sheltered safely during the hours of daylight, and begin to feed (see p. 123). The creatures that now extend innumerable tentacles into the

darkness are mainly invertebrates, such as Feather Stars and Basket Stars, but predominantly myriads of nocturnally active coral polyps. Cardinalfishes and Soldierfishes take the place of diurnal plankton-feeders. It has been observed that even if they were to feed at the same time, they would still not be in direct competition, since they feed at different levels. For example, the Cardinalfishes, *Apogon lachneri* and *A. maculatus,* catch planktonic organisms at about 25 cm above the ground, while *A. townsendi* and *Phaeotery conklini* do so at a height of 2 to 3 m (Luckhurst and Luckhurst, 1978). Observations were carried out off Palau of six species of Sea Cucumber. Three of them fed round the clock, the other three only in the evening and at night. The Holothuroid *Stichopus,* living in the same habitat, feeds only during the hours of daylight. Sea urchins spend the day concealed in holes in the rocks and make their way to the pastures only at dusk. Animals that are found almost exclusively at night include many crustaceans and molluscs, such as crabs, crayfish, shrimps, squid, octopuses, cones and nudibranchs (Ill. 116).

The changeover from "night shift" to "day shift" for the occupants of a coral reef takes place in the very brief dawn and dusk periods. During these few minutes, the reef appears swept bare—a disconcerting sight. Since the one group is already tired and the other not yet fully awake, it is now that the major predators can take their prey most easily. Sharks, Barracudas, Firefishes, Morays and Rock cods go hunting in this, the most dangerous quarter of an hour for many creatures. As the sun rises higher, they disappear (Ill. 119). The many small fishes now leave their refuges, and soon their brilliant colours and diverse forms enliven the reef, after the night-revellers have disappeared into the maze of corals and caves (Fig. 29).

But the diver, exploring at night, may well find the diurnal creatures in deep sleep in caves or under rocks and corals. Fishes that are seen during the day in shoals have dispersed to individual hiding places. Here and there, an isolated Fusilierfish may be seen. Moorish Idols lie resting against rocks, Parrotfishes have retreated into narrow hiding places, Butterflyfishes and Damselfishes have wedged themselves between rocks with their fins. A diver is able to approach very closely and shine his torch straight into their lidless, permanently open eyes. They are not disturbed, indeed, in their somnolent state, they scarcely show the least flicker of movement. Many Parrotfishes sleep inside a fine transparent mucous covering that they have made themselves. When they want to sleep, glands lying beneath the gill covers secrete mucus that is passed out through the gill plates together with the respiratory water, until finally it surrounds the entire fish. An aperture is left at the front and back to ensure a supply of fresh water. The mucous covering prevents the emission of electric impulses and odorous substances from the fish into the water. In this way, they remain undetected by their enemies, the sharks, that are equipped with electro-receptors, and the morays that have an extremely keen sense of smell.

A number of reef creatures are specialists in using stony corals as their homes. The extent of their dependence varies, as does the amount of damage to the host, which in most cases is negligible, but which can occasionally verge on parasitism. Coral colonies, with their complicated branchings and recesses, offer many creatures a safe refuge, with the additional protection of the stinging cells of the polyps. On the other hand, the presence of the cnida restricts the number of coral inhabitants, since only relatively few are immune to the poisons they contain. Permanent inhabitants of stony corals include organisms that run, climb or crawl on hard ground, such as featherstars and brittlestars that are often

no bigger than a penny, the sea urchin *Eucidaris metularis* that browses on corals, a large number of bristle worms, Pistol shrimps (*Alpheus* and others), some 20 genera of astonishingly well-camouflaged true shrimps, vividly coloured or else camouflaged coral crabs, smoothly gleaming Porcelain crabs (Porcellanidae), a few molluscs and fishes (Ill. 88). The nettle cells cannot harm creatures with thick, hard exoskeletons. So crabs and shrimps are common among the branches of corals. In colonies of Styloid coral, *Pocillopora damicornis,* 55 different decapod crabs have been counted. Crabs are skilful climbers. The species *Trapezia* and *Tetralia* that are represented in the largest numbers, live on branching colonies such as Staghorn, Styloid, Claviform and Needle corals. The crabs are highly dependent upon corals, but *Trapezia ferruginea* has been observed to migrate from one colony to another under the cover of darkness.

Many creatures live in the lower parts of branching coral colonies that are usually dead and no longer have batteries of stinging cells. A rich supply of food is available in countless cavities and corners in which detritus has been trapped or has settled. In a single dead colony of *Stylophora*, 1,441 polychaetes (Polychaeta) of 103 species were found. In addition, cheliferous slaters, sea fleas and isopods were common. Sipunculid worms, decapod crabs and brittlestars occured sporadically. Young Morays and Sea Bass also find shelter there.

The protection of live coral colonies is sought by Damselfishes of the genera *Dascyllus, Pomacentrus, Chromis, Abudefduf* and *Glyphidodon* in particular, as well as by bright red Jewelfishes *(Anthias)* (Ill. 32, 87, 92). *Dascyllus* species are dependent upon branching corals, but decreasingly so in the following order, *D. marginatus, D. aruanus* and *D. trimaculatus*. 5 to 30 fishes are found in a colony, depending upon its

28 Characteristic distribution of fishes in an Indo-Pacific coral reef. 1 Fusilier fishes, Pterocaesio diagramma, 2 Jewelfishes, Anthias squammipinnis, 3 Humphead, Bolbometopon muricatus, 4 Manta, Manta birostris, 5 Bullfish, Heniochus monoceros, 6 Six-banded Sergeant Major, Abudefduf sexfasciatus, 7 Shoal of mullets, Valemugil seheli, 8 Goby, Ctenogobius nebulosus, 9 Bothid Flounder, Bothus pantherinus, 10 Wobbegong, Orectolobus wardi, 11 Rainbow wrasse, Thalassoma melanochir, 12 Blue-striped Snapper, Lutjanus kasmira, 13 Butterflyfish, Chaetodon trifasciatus, 14 Grouper, Cephalopholis pachycentron, 15 Blue-lined Surgeonfish, Acanthurus lineatus, 16 Wrasse, Thalassoma lunare, 17 Butterflyfish, Chaetodon lunula, 18 Butterflyfish, Chaetodon auriga, 19 Dark Parrotfish, Callyodon niger, 20 Three-spotted Damselfish, Pomacentrus tripunctatus, 21 Firefish, Pterois volitans, 22 Cardinalfish, Ostorhinchus apogonides, 23 Soldierfish, Holocentrus diadema, 24 Cowfish, Lactoria cornuta, 25 Grouper, Epinephelus salonotus, 26 Wrasse, Coris caudimacula, 27 Toby, Canthigaster valentini, 28 Porcupine fish, Lophodiodon calori, 29 Damselfish, Chromis simulans, 30 Sweetlips, Gaterin gaterinus, 31 Birdfish, Gomphosus caeruleus, 32 Regal Angelfish, Pygoplites diacanthus, 33 Zebra fish, Zebrasoma veliferum, 34 Triggerfish, Melichthys viduz, 35 Moorish Idol, Zanclus cornutus, 36 Red Mullets, Upeneus sulphureus, 37 Jawfish, Gnathypops rosenbergi, 38 Wrasse, Stethojulis axillaris, 39 Surgeonfish, Acanthurus triostegus, 40 Lizardfish, Synodus variegatus, 41 Stonefish, Synanceichthys verrucosus, 42 Parrotfish, Callyodon africanus, 43 Three-striped Damselfish, Dascyllus aruanus, 44 Spotted Ray, Taeniura lymna, 45 Butterflyfish, Chaetodon lineolatus, 46 Hawkfish, Paracirrhites forsteri, 47 Butterflyfish, Chaetodon falcula, 48 Wrasse, Halichoëres centriquadrus, 49 Angelfish, Pomacanthops filamentosus, 50 Shark, Carcharhinus spallanzi, 51 Unicorn fish, Naso unicornis, 52 Copper snapper, Lutjanus bohar, 53 Sling-Jaw, Epibulus insidiator, 54 Emperor Angelfish, Pomacanthodes imperator, 55 Moray, Lycodontis undulatus, 56 Grouper, Epinephelus tukula, 57 Nurse shark, Ginglymostoma brevicaudatum, 58 Tube eels, Xarifania sp. (Original)

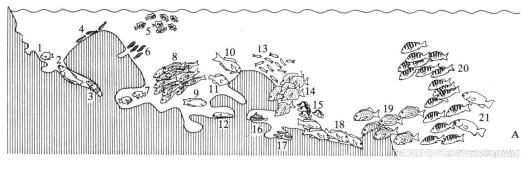

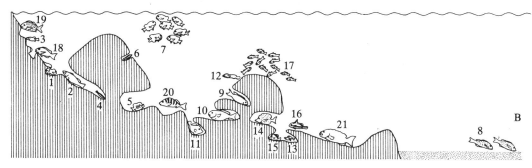

29 *Fishes in a Southern Californian reef (A) during the day and (B) at night.*
1 Eupomacentrus rectifraenum *(Damselfish)*, *2* Epinephelus labriformis *(Grouper)*, *3* Holocentrus suborbitalis *(Squirrelfish)*, *4* Thalassoma lucasanum *(Wrasse)*, *5* Abudefduf troschelii *(Abudefduf)*, *6* Runula azalea, *7* Myripristis occidentalis *(Sol-dierfish)*, *8* Microlepidotus inornatus *(Snapper)*, *9* Bodianus diplotaenia *(Hogfish)*, *10* Scarus cali-forniensis *(Parrotfish)*, *11* Balistes verrea *(Trigger-fish)*, *12* Rypticus bicolor *(Soapfish)*, *13* Chromis atrilobata *(Damselfish)*, *14* Prionus puncataus *(Saw-tai)*, *15* Heniochus nigrirostris *(Bullfish)*, *16* Pareques viola, *17* Apogon retrosella *(Cardinal-fish)*, *18* Lutjanus argentiventris *(Silver-Snapper)*, *19* Anisotremus interruptus *(Margate)*, *20* Haem-ulon sexfasciatum *(Grunt)*, *21* Mycteroperca rosa-cea *(Grouper) (After Hobson, 1965)*

size. During the day, they swim above the coral that is their home, snapping up plankton, but all the while remaining so close to it that they can retreat immediately. If a diver approaches the shoal slowly, the fishes draw closer to their hiding place in direct proportion to the decrease in distance between them and the intruder. The moment the minimum safety distance is exceeded, all the *Dascyllus* disappear instantly among the coral branches, each one to a particular place that is exclusively its own. Jewel-fishes that often occur in shoals of several hundred individuals, behave in a similar way. They dart among the coral boulders in such a lively way that they resemble a shower of sparks. When danger threatens, each fish disappears rapidly into one of the many holes or cracks that a colony must have if it is to serve them as a home.

Anemone- or Clownfishes *(Amphip-rion)* have evolved the ability to hide from danger among the tentacles of the actinians *Radianthus* and *Discosoma*, that otherwise deliver a fatal sting (Ill. 125). They too hover above the ane-mone in which they live, catching plankton. In terms of behaviour, the Three-spot Humbug *Dascyllus trimacu-latus* holds a position midway between that of the *Dascyllus* and *Amphiprion* representatives. As a young fish, it finds protection among the tentacles of sea anemones, when sexually mature, among stony corals. Is it on the way to becoming a "convergent Anemonefish"?

Conspicuous in the coral reef are the Hawkfishes *(Cirrithidae)*, small fishes finger-length to a hand's span, with a club-shaped body and vivid markings. They rest on a coral branch utterly motionless except for the protruding eyes that roll vigilantly from side to side (Ill. 36). Approached too closely, they retreat among the branches of the coral or move on to the next colony. More strongly dependent upon corals for their homes are some of the gobies (Gobiidae) of the genus *Paragobiodon*. In densely branching colonies of *Madracis mirabilis*, 19 species of fish were found, predomi-nantly Gobies and Dwarf Sea Bass (Grammistidae).

Many minute cyclopes (Copepoda) are specialist residents in corals, and it was only in the last few years that they were discovered and described (Fig. 30). Snails are also among the "coral loving" ani-mals, and appropriately, one family, the members of which occur exclusively on stony corals, Black corals and Gorgo-nians is called *Coralliophilidae* (Lat. Gr., coral loving). Other inhabitants of Scler-actinia include Architectonicidae, Epito-niidae, the muricide *Drupa*, the trochide Calliostoma and a number of Egg cow-ries (Ovulidae) and Sea Slugs (Nudibran-chia).

Other animals grow firmly attached to stony coral colonies. Most of them are mussels of the genus *Electroma* on Staghorn corals, the oyster *Ostrea sand-vichiensis* on Styloid coral and *Barbatia decussata* on Pore corals. They colonize only those corals that grow to a height sufficient to enable them to avoid silta-tion. Mussels known as Watering Pots (Clavagellidae) have much reduced shells, that have been replaced by a sec-ondary calcareous integument which in some species is anchored between corals. The forward surface is closed off by a perforated plate. Worm-shells or "Old Maid's Curls" (Vermetidae) are not host-specific and are attached to various species of coral, even to an inanimate, hard soil substrate. Sometimes they allow themselves to be enclosed in a growth of coral. On stony corals, calcified tubes that have become empty because the creature living there has died, are immediately taken over by others. Vacated tubes made by the worm *Spirobranchus giganteus* (Ill. 60, 101) are taken over by the Hermit crab *Paguritta harmsi,* which thereafter lives a sedentary life, having successfully occupied a new ecological niche.

An even closer association with live stony corals is that shown by organisms living inside it. Either they bore into the coral or allow it to grow round them. The latter process is less harmful to the corals. Most of the organisms are bivalves, including the Sea Date. The minute *Fungiacava elatensis* that is found in the Gulf of Aqaba, lives, as its name suggests, in small holes in Mushroom corals. As a result of enclosing growth, characteristic skeletal excrescences may develop, which, by analogy with similar structures on plants, may be called "coral galls". *Pedum spondyloideum,* a relation of the Scallops, attaches itself firmly to stony corals with byssal filaments while still in its juvenile stage, and, in time, allows the coral to grow round it. The hole leading to the outside remains large enough for the delicate, iridescent shell to be opened to a width of about one centimetre. Snails of the genera *Magilus* and *Leptoconchus* live embedded in scleractinian skeletons. The increase in size of their tubular shells is in proportion to the growth of the host coral. Sabellid worms also allow coral to grow round them. Some of them appear to be host-specific. In Blue corals, a Polychaet, probably of the species *Leucodora,* makes minute holes and passages. Tubicolar annelids inhabit *Mycedium, Porites* and other stony corals. On approaching Pore and Fire corals in calm water, a diver will frequently see rings of tentacles, like feathery spirals,

measuring 1–2 cm, brightly coloured in white, yellow, red, blue or patterned in two colours, which are the front ends of Spiral worms of the cosmopolitan genus *Spirobranchus*. At the least disturbance, they disappear in a flash into the tubes in which they live, plugging up the hole with a lid-like plate.

Among crustaceans, many sessile balanids allow themselves to be enclosed in coral growth (Ill. 91). Observations of *Pyrgoma* have shown that the larva settles without being attacked by the polyp's stinging cells, destroys the living tissue and within two days, reaches the skeleton to which it attaches itself head downwards. After that, the process of metamorphosis into the balanid form begins, and coral grows, enclosing it in a cone with a small opening at the top. In all, species from about a dozen balanid genera live embedded in stony corals, fire corals and calcareous algae. But certain free-moving crustacean groups such as the shrimp *Paratypton siebenrocki* and a number of crabs of the genera *Troglocarcinus, Cryptochirus* and *Hapalocarcinus* also allow corals to grow and surround them (Fig. 30). *Troglocarcinus,* a specialist of Mushroom corals, becomes embedded in the thickened, bulging septa of the skeleton. *Cryptochirus* in its cylindrical form, about 1 cm long, lives in tube-shaped holes in massive stony corals. It can close the aperture with its cephalothoracical plate.

Hapalocarcinus lives on Styloid, Claviform and Pore corals, where the females produce striking pouch-like galls at the ends of branches, in which they are enclosed. The males live freely and are so minute that they can make their way to the females through small apertures, in order to mate with them.

Like stony corals, many other sessile organisms in reefs provide accommodation for a wealth of other species of animals, that is again utilized to the full. For example, the limestone tubes of worm molluscs (Solenogastres), that are

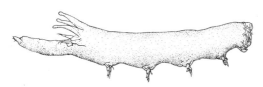

30   Crustaceans in corals. The small crab, Pseudocryptochirus crescentus *(top)* allows stony corals to grow up round it and so leads a protected existence *(after Jarth and Hopkins, 1968)* while the Sea Flea, Xarifia lamellispinosa, *living as a parasite in coral polyps, no longer needs shell and legs, and consequently they have regressed. (After Patton, 1976)*

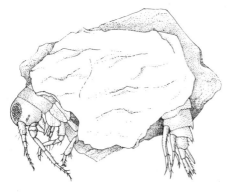

31   The coral sand between the reefs of Barbados is inhabited by the small Sea Flea, Tiron bellairsi, *which uses its leg claws to hold a small disc of coral sand against its body on either side as a protective shield. (After Just, 1981)*

about 0.5 to 3 cm long, are found in branching colonies of hydrozoa, Gorgonian and Soft corals. Sometimes they have entwined the colony, growing round it until they are inextricably linked, sometimes they are capable of crawling by means of utricular musculature or of a ventrally situated ciliated furrow. Some of them have a horny, rasping tongue, like the radula of a snail. We are not able here to describe in detail the complicated relationship that exists between host and associated organism. But we shall return later to a consideration of some common and characteristic examples of symbiosis.

Locomotion within a reef

In accordance with their locomotive capacity, it is possible to distinguish between organisms that are fundamentally sessile or sedentary, those that are hemisessile or scarcely motile, and those that are vagile or errant. In a coral reef, benthic organisms (see p. 94) predominate, whether they are sessile, hemisessile or vagile. We have already come across many sessile organisms, in particular corals. Also included here are solitary creatures that have grown into or onto a substrate, such as many bivalves, snails, crabs, balanids (Cirripedia) and a related species, the stalked Goose Barnacle (also known as Gooseneck Barnacle) *(Lepas)* (Ill. 94). Apart from algae and, in a sense, sponges, all of them are equipped with an innervated muscular system, and in a variety of ways, are capable of movement, enabling them to react to danger and to gather food.

The life of hemisessile organisms on or in the substratum only appears to be stationary. The ascidian, *Didemnium molle* and related species, alter their position only by millimetres in the course of days. It may be that because they too live in symbiotic association with algae, they are following the incidence of light, which, even in the tropics, alters slightly in the course of the year. Sea anemones push themselves slowly across the hard ground with undulating movements of their underside. Some detach themselves from the base, contract into a sphere and allow themselves to be rolled along by the movement of the water. The small sea anemone *Boloceroides mcmurrichi,* found in the Red Sea, pumps itself full of water, spreads its tentacles and, weightless, allows itself to drift. By rhythmical beating of the tentacles, it is able to cover considerable distances (Ill. 99). Crawling bivalves extend their siphons up out of the sand. If the habitat loses its appeal, they extend their muscular, tongue-shaped or worm-like foot through the shell aperture into the sand, cause it to thicken at the end by congestion of the blood vessels, and then draw the body and shell across it. In this way, they plough laboriously through the ground. Sessile bivalves, on the other hand, anchor themselves firmly to a solid base by means of tough byssus filaments, or they cement the lower half of the shell to rock or wood; oysters, in which the foot is now merely vestigial, are an example.

Vagile animals live on the ground or freely in the water. Vagibenthic organisms exhibit a wide variety of different forms of locomotion. Snails and flatworms, sea urchins, sea stars and sea cucumbers are crawlers. They slide along in close contact with the ground, using different means of locomotion. Snails, for example, send waves of contractions along the foot, flatworms perform synchronized movements of the ciliated epithelium (Ill. 113). Starfishes, sea urchins and sea cucumbers exhibit a mode of locomotion unique in the animal kingdom, the ambulacral system. Internally, they have a system of branching water channels extending radially from a central ring canal. Arising in the radial channels, hundreds of tube feet, arranged in rows, protrude through small apertures in the calcareous skeleton. As the pressure of water inside them is increased, the suction discs curve outwards, the feet relinquish contact with the base, perform a coordinated movement forward in the direction of travel and draw the starfish on. With the relaxation of pressure, they once again adhere firmly to the ground. Species with conical feet that do not attach themselves by suction, such as the Burrowing starfish *(Astropecten),* live on sandy ground in reefs and lagoons. A large number of crawlers are represented by annelids, which raise the body slightly from the ground by means of stub feet (parapodia), at the same time performing undulating motions. Countless numbers of crustaceans, some only just visible, some up to finger-nail size, such as ostracods, sea fleas and isopods also move across the ground by crawling. Bepincered cheliferous slaters (Tanaidacea) crawl about in the protective tubes they themselves have spun and anchored among algae. They are more common in the adjacent mangrove zone than in the reef. Creatures that run or walk include ground-dwelling crustaceans—brachyurans, crayfish, thalassinids (Thalassinidea), Slipper lobsters (Scyllaridae) and many shrimps. Strong paired legs lift the body above the ground and carry out rapid coordinated movements. The Arrow crab *(Stenorhynchus),* a curious creature with a small body, long nasal process and thin legs, is a real "stilt walker". There are also some very specialized fishes such as Batfishes (Ongcocephalidae) and Gurnards or Sea Robins (Triglidae), in which the pectoral fins are developed into long, powerful filaments which they move like legs to walk across the ground. Similarly, Brittlestars, when in retreat, raise the small disc of a body on arms and "run" along on the rapidly waving arms. All creeping, crawling and running organisms are in general also skilled in climbing.

Those species adapted for digging, the fossorial organisms (L. *fossa-fodere*, to dig), can be found in large numbers in the sandy areas that always exist in a reef region. They are sipunculids, ringed worms and acorn worms (Enteropneusta), crustaceans, snails, bivalves, sea urchins and even fishes. They have adapted to this way of life, and in many cases exhibit appropriate features. The acorn worm has a smooth, rounded, hard head with which it can easily penetrate fine sand and silt. Locomotion is achieved by muscular distension and contraction, while a ciliated epithelium conveys loose particles of sand backwards and to the mouth. Burrowing sea urchins have elongated, much flattened bodies with short, dense, flexible spines with which the creature performs rotary or spiral movements that push it through the sand. The Sand Dollar *(Melitta)* has evolved a completely discoid form that offers minimum resistance. Many crustaceans have extremities with a shovel-shaped extension, Burrowing crabs (Hippidae) have caudal plates in the form of lancet-shaped digging implements, Tube-dwelling eels a-burrowing tail. Moon snails (Naticidae) inflate and deflate their foot and in this way, force themselves through soft ground with only the breathing tube protruding like a snorkel. Particular species are often adapted to ground with a particular size of particle.

Animals of pelagic habitat can be divided into two groups; firstly, those that in spite of a degree of independent movement are subject to the movement of the water, that is, they float or drift. These are planktonts in the narrow sense. The second group consists of swimmers or nektonic organisms, the independent movements of which prevail against those of the water, and which are known collectively as nekton (Gr. *nektos,* swimming).

Plankton includes myriads of microscopically small organisms in suspension—flagellate algae, foraminifers and radiolarians, cyclopes, sea fleas and mysid shrimps as well as large numbers of larvae of sessile organisms. Also included are siphonophores, medusae, comb jellies or ctenophores and pyrosomes that can grow to a considerable size. Why do planktonic organisms float in suspension? Their specific gravity is always slightly greater than that of water, but they reduce or compensate for sinking by various physical devices. Skeletons, such as those of Peridiniacid algae and diatoms are fragile and of light construction. Frequently, body processes extend the surface area. They increase friction to such an extent that the sinking rate is almost nil. And buoyancy can be increased by the presence of fatty droplets stored inside the body as reserves or by abundant gelatinous tissue, by body fluids with a low salt content or by gas bubbles. Hydrodynamic effects and the creature's own movements entirely prevent sinking. Motile planktonic organisms such as cyclopes (Copepoda) and sea fleas (Cladocera) can even move actively in a vertical direction. At night, they rise in dense clouds to the surface of the water and in the morning, allow themselves to sink down again.

Nektonic organisms are represented primarily by fishes and cephalopods (cuttlefishes etc). Depending upon size, shape of body, habitat, territory and feeding habits, the fishes of the coral reefs have evolved a variety of swimming techniques. Rapid, rectilinear, tail-driven locomotion that predominates among fishes of the open sea and those that swim for long distances, is found in only relatively few species such as the torpedo-shaped to spindle-shaped Barracudas *(Sphyraena),* Trumpetfishes *(Aulostoma)* and Flutefishes *(Fistularia),* as well as Fusilierfishes *(Caesio)* that swim in shoals (Ill. 110). Many other fishes, such as Wrasses, Butterfly- and Angelfishes, Sweetlips, Grunters (Haemulidae), Damselfishes and Rock cods also propel themselves by movements of the caudal fins, but in all these cases, there are additional movements carried out by the dorsal, anal, ventral and pectoral fins, which considerably enhance manoeuvrability, as do the tail units, ailerons and landing flaps of an aeroplane (Ill. 83, 84, 87, 92, 93, 106, 112).

Pronounced use of the pectoral fins in locomotion is characteristic of Surgeon- and Parrotfishes. Puffers and Trunk fishes whirl their pectoral, ventral and anal fins. Although they are slow and have little endurance, they are as versatile in movement as helicopters and skilled in reverse manoeuvring (Ill. 108, 117, 127). Triggerfishes execute sideways strokes of the dorsal and anal fins in opposed directions (Ill. 107). Morays, Sand sharks and Reef sharks as well as minute Gudgeons *(Nemateleotris)* that hang motionless outside their holes, move rapidly and dexterously through the reef with powerful serpentine movements.

Flatfishes and Sting Rays are gliders. With their undulating continuous fin, they create a propulsive thrust and then glide along. Manta rays (known as Devilfishes) that can have a span of 6 m or more, and Eagle rays raise and lower the pointed fins that extend far out on either side, so that their motor pattern resembles avian flight. Sharks also glide. Their hydrodynamically efficient body-shape and the sculpted pectoral fins that are spread sideways like aeroplane wings produce a favourable gliding angle. In this way, they compensate for the greater expenditure of energy required of them for swimming and hovering, in comparison with those fishes equipped with air bladders.

Some of the slowest swimmers are probably the Sea Horses that live among sea fans and sea grasses; in flight from an enemy, they may well exert themselves and progress slightly more rapidly by using the dorsal and pectoral fins, but when undisturbed, they maintain an upright position and move only millimetres at a time. To this end, water taken in for respiration is forced out through the gill covers that are moved up and down, thereby achieving a minute propulsive thrust. The Batfish also uses the principle of retro-thrust to dart away when danger threatens. But the water is forced out through nozzle apertures situated behind the pectoral fins that are extended like arms, instead of through gill covers. Occasionally, groups of nektonic cephalopods, sepias or squids occur in the reef. Usually they swim with the continuous fin held back, but when an enemy approaches, they disperse in a flash like tiny rockets with reaction propulsion. Ground fishes such as Scorpionfishes, Firefishes, Stonefishes, Flatfishes, Gudgeons and Blennies move forward in short darts, driven by the action of the fins.

Almost all the fishes living in the highly fragmented biotope of the coral reef exhibit very mobile eyes. Using them, they follow every tiny alteration and movement in their surroundings. It has been shown experimentally that they also have a very good memory for places and, in contrast to fishes of the open sea that live in shoals, such as the staring-eyed herring, a comparatively high level of intelligence. Because of the complex nature of their habitat and the large numbers of creatures living there, they have to distinguish a wide variety of environmental stimuli and react rapidly to them; so it seems plausible to regard this development as a result of environmental influences.

Feeding habits in the coral reef

Food is the basic element of metabolism, growth and reproduction. It is as widely varied as the animal world itself (Fig. 32), and nowhere is the relationship between organisms as close as in this sphere: one organism consumes another. But instead of existing in a state of deadly rivalry, the coral reef complex is a balanced one, because every species exhibits a different, though sometimes only slightly divergent feeding behaviour.

The waters of the ocean contain all the chemical elements in solution, although usually only in traces. Plants—unicellular algae that occur in vast numbers and often in curious forms among the plankton, and even affect the colour of the water, multi-cellular green, brown or red algae as sedentary compact masses, sheafs, gelatinous balls, membranous lobes, leathery straps, plump berries and sturdy stems with leaf-like appendages or simply as blackish-blue encrustations, and various species of sea grass—assimilate inorganic matter using the energy of the sun. Phosphorus is dissolved in the sea in such small quantities that the production of phytoplankton is restricted. In addition, a considerable amount of nitrogen is consumed in reef regions because, apart from free-living algae, the zooxanthellae of the corals, flatworms and molluscs absorb nitrates as nutritive salts. Even though every day a myriad of planktonic gametes, spores, eggs, spermatozoa and larvae is released, the water usually becomes slightly cloudy only at night, when organisms in suspension rise from the depths towards the surface. Since they sink again with the arrival of daylight, and as a result of the intensive catching and filtering activity of the many plankton consumers, the water above the reef becomes clear again after sunrise, so the assimilation process is not impeded.

In addition to inorganic salts, sea water contains organic compounds in solution, of which many are suitable as food for various organisms. Mention has already been made of their osmotic absorption (see p. 24).

Another source of food in the reef is

113 82

82 *Squirrelfishes* (Adioryx *sp.*) *are nocturnal. By day, they retire into dark crevices and caves. (Society Islands, Pacific)*

114/115 83 84 | 87
     85 86 |

83 *The usual resting place of Sweetlips* (Gaterin gaterinus) *is beneath rocky overhangs or in the protection of corals. (Red Sea)*

84 *A Regal Angelfish* (Pygoplites diacanthus) *emerges from a narrow gulley that is surrounded by small Pore corals* (Synaraea irregularis) *in both the leaf-like and stumpy, branching growth-forms. (Society Islands, Pacific)*

85 *The Pacific Bullfish* (Heniochus chrysostomus) *usually remains close to hiding places into which it can escape if danger threatens. (Society Islands, Pacific)*

86 *Sergeant-major fishes* (Abudefduf sexfasciatus) *that feed on plankton in open water above the reef shortly after sunrise, spend the day hiding in holes or close to them. (Red Sea)*

87 *The number of Yellowtailed Damselfishes* (Dascyllus flavicaudus) *living in a coral, here a Claviform coral* (Pocillopora verrucosa), *depends upon its size. (Society Islands, Pacific)*

116 88 |

88 *Every stony coral has its residents. All the crustaceans, molluscs and ringed worms shown, lived in the branching colony of Styloid coral* (Stylophora pistillata). *(Red Sea)*

89 *The Boring urchin* Echinometra lucunter)
*drills out suitably sized hollows in rocks in which to
protect itself from the pounding surf. (Virgin Islands,
Western Atlantic)*

90 *Whole corals that are fused into the limestone
of the reef with structural details well preserved, like
this Brain coral* (Diploria strigosa), *are not very
common, since the skeletons are usually pulverized
by the action of the waves. (Cuba, Windward
Passage)*

91 *The larvae of highly specialized balanids have
attached themselves to a Staghorn coral* (Acropora
prolifera), *stimulating the life tissue in such a way
that a raised "coral gall" has grown round them,
facilitating the capture of plankton. (Southern
Florida, Florida Strait)*

92 *The Damselfish* Plectroglyphidodon dickii *lives
a solitary life among corals. (Society Islands,
Pacific)*

93 *Butterflyfishes* (Chaetodon pelewensis) *feel
completely safe among the branches of coral, where
they also find their food. (Tuamotu Archipelago,
Pacific)*

the abundant bacterial flora. In addition to corals, certain other water-propelling and filtering organisms such as sponges (Ill. 103–105), bryozoa, bivalves and ascidians certainly make use of it. But more than anything else, microscopically small organisms feed on it. Unicellular rhizopods surround the bacteria with their plasmic pseudopods and digest them rapidly. Infusoria and rotifers direct them by the beating action of cilia towards the mouth. If clouds of plankton drift into the reef at night, most of the sessile and hemisessile creatures become active. Zoantharians (Ill. 98), Gorgonians and soft corals and hydrocoralls extend their tentacles into the stream. All of them catch plankton. As soon as a zooplanktonic organism touches the tentacles, it is paralyzed by injections of nettle-cell poison and directed into the digestive cavity (see p. 23). Plankton-eating organisms that have no nettle cells, acquire their food in a different way. Filtering and netting adepts sift drifting organisms from the water with finely reticulated or feathered arms and tentacles with varying sizes of mesh. The bush-like branching tentacles of the climbing Sea Gherkins *(Cucumaria, Bohadschia)* and of the Worm Cucumber *(Synapta)* secrete a sticky substance to which plankton adheres (Ill. 102). At fairly regular intervals, they slowly fold in one tentacle after the other, insert it into the mouth and scrape the food from it. At night, Feather Stars and Basket Stars crawl and roll to the highest of their feeding places. The Feather Stars spread their pinnatipartite arms diagonally sideways (Ill. 49, 50). A food groove runs along the central axis of each one, and food entrapped upon it is transported by the beating action of cilia to the mouth, as if on a conveyor belt. Basket Stars extend their branching tentacles over a distance of up to 70 cm like drag-nets. The meshwork consists of many flexible hooklets and curled filaments. If floating matter becomes

entangled in them, it is held fast. Once the catching device is full, the arms are folded inwards in turn and the food scraped off in the mouth. In calm, sun-warmed lagoon waters, brownish-yellow medusae of the genus *Cassiopeia* lie with the umbrella on the sandy ground. They too house large numbers of unicellular symbiotic algae, and in this way, provide them with optimal conditions for assimilation. In addition, the hemisessile jellyfishes catch sinking planktonic organisms in the tentacles that they stretch upwards. Tube-dwelling worms (Serpulimorpha) extend fine, pinnate circles of tentacles into the water to catch plankton. Sessile crabs obtain their food by filtration. The hermit crab *Paguritta harmsi* that inhabits deserted tubes, catches plankton with its large antennae that, in their morphology, resemble the feelers of male bombyx.

Organisms that capture food by the use of mucilaginous filaments live a sessile life on or in the ground, and extend long, sticky tentacles, mucus-covered filaments or nets. For example, Polychaetes of the Terebellidae family, such as *Eupolymnia nebulosa* explore the surface of the ground with threadlike head tentacles. Drifting food particles sinking down onto them are held fast, to be conveyed to the mouth by retraction of the tentacle. The whitish feeding tentacles of *Reterebella queenslandica* are almost one metre long. The tube-dwelling worm-snails known as Old Maid's Curls *(Vermetus)* throw out mucus-covered filaments up to 40 cm long in a sudden movement when the water is agitated by the presence of planktonic organisms within a radius of up to 20 cm. If plankton adheres to them, the snail retracts the filaments with its radula and draws the prey into its mouth. There are creeping comb jellies (Ctenophora) on the spines of sea urchins, under the umbrellas of great soft corals *(Sarcophyton)* or on the ground, which at night extend thin, sticky tentacles of up to

32 Animals living on or near a coral colony have different ways of feeding.

I. Coral feeders: *1 Porcupine fish* (Diodon), *2 Parrotfish* (Callyodon).— II. Plankton feeders: *3 Featherstar* (Comatula), *4 Damselfish* (Dascyllus), *5 Jewelfish* (Anthias), *6 boring snail* (Leptoconcha), *7 boring sponge* (Cliona), *8 boring mussel* (Lithodomus), *9 boring crustacean* (Cryptochirus), *10 Barnacles* (Pyrgoma), *11 Chalk-tube worm* (Serpula), *12 tunicates* (Botryllus), *13 Moss animalcules* (Bryozoa), *14 Ghost shrimp* (Caprella), *15 soft coral* (Lobophytum), *16 sea anemone* (Actiniaria), *17 sponges* (Demospongiae), *18 Giant clam* (Tridacna).— III. Animals that feed on small live food and detritus: *19 Gobies* (Gobiidae), *20 Coral crab* (Trapezia), *21 Butterflyfish* (Chaetodon), *22 Longnose Butterflyfish* (Chelmon), *23 Pistol shrimp* (Alpheus), *24 Sluggish fish* (Caracanthus), *25 Bristleworm* (Nerëidae), *26 Brittlestar* (Ophiuroidea), *27 Sea Cucumber* (Cucumaria), *28 snail* (Gastropoda), *29 Sea urchin* (Echinometra), *30 Amphipod* (Amphipoda),— IV. Predators: *31 Cone* (Conus), *32 Cushion starfish* (Pentaceraster), *33 Flatworm* (Polycladida), *34 Naked-gilled sea-slugs* (Nudibranchia), *35 Moray* (Gymnothorax), *36 Grouper* (Epinephelus)
(After Gerlach, 1964, modified)

122

1.5 m length into the streaming water. If floating matter becomes entangled in the adhesive cells, the tentacle is retracted and the food scraped off in the mouth.

Cirripeds (Cirripedia) of the genera *Tetraclita* and *Kochlorine,* that grow freely or enclosed within corals or calcareous algae, sweep the water rhythmically with pinnatipartite tentacles that are extended through an opening to filter plankton from the sea. Since in doing this, they create a slight current of water, they form a transition to another method of feeding on plankton, and to those organisms that feed by means of water currents.

These organisms do not rely on plankton drifting passively towards them, but create a directed water current from which they extract food organisms and which, at the same time, provides fresh, oxygen-rich water for respiration. Within the body-wall of a sponge, there are innumerable microscopic flagellated chambers that are linked to the outside water through pores. The activity of the flagella creates a continuous incurrent flow of water. Fine organisms are retained and digested, while water is swept away through the central osculum. A sponge with 20 finger-shaped protuberances 10 cm long and with a diameter of 2 cm, filters 1,575 litres of water in one day. This remarkable capacity demonstrates the important part sponges play in maintaining the clarity and purity of water in a coral reef. Bivalves that grow attached to rock faces or among corals—Thorny Oysters *(Spondylus),* Pearl Oysters *(Pteria margaritifera),* Noah's Ark shells *(Arca)* and Giant Clams *(Tridacna)*—vibrate their ciliated epithelium to convey plankton to the mouth through intake apertures, and at the same time, fresh water round the gill lamellae. Sessile, cylindroid ascidians or Sea Squirts use ciliary action to direct a stream of water through the highly perforated gill pouch with which they filter out planktonic organisms. In minute bry-

ozoa tentacular cilia convey the finest nanoplankton, consisting of bacteria and protozoa, into the body cavity. The small crabs *Hapalocarcinus marsupialis, Cryptochirus coralliodytes* and *Troglocarcinus fagei* and related species, that live in coral galls, create a stream of water with their extremities and oral limbs that carries food to them. Other organisms that feed by means of water currents are Porcelain crabs (Porcellanidae), common in closely-branching corals. They extend the pinnate bristles on the oral limbs to form a spoon-shaped fan, and by folding and unfolding it, continually force water through the meshes of the net. Mouth parts comb out plankton that has been caught. By regular vibration of the antennae, the Hermit crab, *Carcinus verrilli,* creates a current of water directed to its mouth, from which it extracts particles of food.

From time to time, giant pelagic fishes swim into the reef areas: they could be described as "basket fishers". Manta rays, for example, are sometimes seen on reef slopes. Like the great whales, these, the largest of all the rays that may weigh a ton, are also plankton feeders. As they move actively forwards, they take in plankton passively. Their wide jaws are almost always open. The jaws are flanked by two wide "horns" or head-fins that direct plankton into the mouth. It is filtered out in a narrow-mesh basket-like apparatus.

Many small fishes are active plankton feeders. They locate the suspended organisms individually and snap them out of the water. They include many Damselfishes, Jewelfishes, (Ill. 109), Butterflyfishes, Fusiliers and others. The hemisessile tube-dwelling eels hold a special position. Stretching high, they look out from the tubes in which they live and snap at edible suspended matter with unerring precision.

The grazers in the coral reefs eat and gnaw at organisms growing on the ground. Some of them are specialized for

biting through tough sea grasses. These herbivores include the Long-spined Hatpin urchins *(Diadema)* with their strong, sharp teeth. During the day, assemblies of these sea urchins, often in large numbers, seek the protection of corals and reefs lying close to sea meadows. At night, they graze on sea grass, so that badly denuded zones two to ten metres wide develop round reefs. Sea urchins of the genera *Psammechinus* and *Strongylocentrotus* also graze on sea grasses, as do certain snails with their sharp, rasping radulae. Rabbitfishes *(Siganus),* common in those parts of the reef close to lagoons, together with Sea Turtles and Sirenians or Sea Cows, that have been overhunted by man and are threatened by extinction, are purely vegetarian, feeding on either seagrass or algae, according to availability.

Many inhabitants of the reef eat algae. Depending upon the supply of coarse seaweeds, fine filamentous or encrusting algae closely covering the rocks, various specialist feeders have developed among them. Seaweed feeders are relatively rare, the major representatives being snails, such as the Abalones *(Haliotis)* or else sea urchins. Soft filamentous algae are the preferred food of many reef creatures, such as the majority of Polychaetes. The horny radulae of snails, that have a wide variety of complicated forms, are particularly suitable for cropping short algae beds and encrusting algae. Food-getting habits therefore vary between species. Large Fighting conchs (Strombidae), Top shells (Trochidae) and Spider conchs *(Lambis),* as well as small Periwinkles *(Littorina)* and Limpets or Tent shells (Patellacea) are among the alga feeders. The latter scrape layers of blue-green algae or diatomaceous algae no more than tenths of a millimetre thick from the substratum. Short-spined sea urchins *(Arbacia, Paracentrotus, Echinometra),* Pencil urchin *(Heterocentrotus)* and the Lance urchin (Phyllacanthus) that live in large numbers in holes in the rocks at the margins of the reef, graze nocturnally. Of fishes that feed on algae, Surgeonfishes (Acanthuridae) graze on the ground in large herds with the head directed diagonally downwards (Ill. 108). In contrast to them, alga-feeding Damselfishes of the genera *Pomacentrus* and *Abudefduf* defend their territory vigorously. In their custody, the alga beds grow more luxuriantly because they are protected from other grazers. But the latter, as they graze, keep the rock substrate open for colonization by corals and calcareous algae, and therefore contribute to the growth and consolidation of the reef. Some of the smaller Parrotfishes in particular, scraping the rock clear with their powerful teeth, assist the process. As they crop their plant food, all the herbivorous grazers ingest animal food at the same time, for there is no blade of sea grass without its growth of hydrozoa, bryozoa, bivalves or worms. And tufts of filamentous algae house large numbers of protozoa, rotifers, threadworms and ringed worms, small crustaceans and snails. All of them are consumed as an incidental extra.

Carnivorous grazers are usually even more highly specialized than herbivorous. They graze on sponges, hydroid polyps, bryozoa, soft or stony corals. Specialists of this kind are found among snails, particularly the nudibranchs, among echinoderms and fishes. Certain Porcelain snails and Angelfishes such as the Hawaiian Angelfish *(Apolemichthys arcuatus),* the Queen Angelfish *(Holocanthus ciliaris)* and the Emperor Angelfish *Pomacanthus arcuatus)* specialize in a diet of sponges (Ill. 136). The "Flamingo Tongue" *(Cyphoma gibbosum),* a snail common in Caribbean reefs, grazes on Sea Fans, *Heliacus,* a snail with a flatly coiled shell, eats the polyps of soft corals. A number of Wormsnails without shells evert the foregut like a proboscis, press the horny teeth of their tongue against a hydrozoan, sea whip or soft coral and tear out particles of tissue.

In spite of batteries of stinging cells, a number of grazers are able to feed on stony corals. They have evolved effective counter-measures in the form of rigid protective integuments and thick skins that prevent the nettle poison from penetrating the body, and mucus that suppresses the explosive discharge of the stinging capsules. Nudibranchs (Fig. 46; Ill. 141) of the genus *Phestilla* are highly specialized; for example, *P. melanobrachia* grazes on ahermatypic Tree corals, *P. sibogae,* on the other hand, on hermatypic Pore corals. The shell-bearing snails of the genus *Philippia,* known as Mediterranean Sun-dials, Rock shells of the genus *Drupa,* Top snails of the genus *Calliostoma, Jenneria* that is related to the Porcelain snail and various Great Egg-shells (Ovulidae) do not harm Scleractinia excessively, but the coral-eating *Coralliophila abbreviata* undoubtedly contributed to the destruction of the Small Button coral, *Montastrea annularis,* in the reefs of Barbados.

The number of Polychaetes that specialize in a diet of stony corals is certainly considerably greater than one might imagine; many of them have surprisingly strong, sharp and pointed mouth-parts. The bristle worm, *Hermodice carunculata,* bites away the polyps of Pore coral, *Porites porites,* and of Staghorn coral, *Acropora prolifera,* to such an extent that the branch ends are denuded of living tissue. For anyone observing a ringed worm with its front end inserted into a coral cup, it is often difficult to determine which is eating which. Many crabs that live among and in coral colonies, also feed on them, although not always on the live tissue. It is generally believed that at night, the Caribbean Long-spined Hatpin urchin *(Diadema antillarum)* feeds not only in seagrass meadows but also on Staghorn corals. Although starfishes of the genus *Culcita* (Ill. 124) ingest animal growth from sea grass and rocks and graze on sponges, they are also prepared to eat

Scalpal corals and Coarse Pore corals. The starfish *Choriaster granulatus* probably also feeds on stony corals.

As a grazing animal, the Crown-of-thorns starfish *(Acanthaster planci)* can cause serious damage (Ill. 118), and newspapers frequently report the havoc wrought by its heavy predation. In 1963, these normally rare starfishes were observed in great numbers by tourists near Cairns in the Great Barrier Reef. Spines with three-edged and razor-sharp tips that cover the many arms of these large starfishes deliver a highly toxic venom and provide an effective defence. To feed, the creatures evert the soft, flexible stomach and secrete digestive ferments onto the colonies of stony corals, attacking in particular *Goniastrea, Turbinaria, Montipora* and the table-shaped Acroporans (Ormond et al., 1973). The beating action of many minute cilia conveys the coral tissue that is now in the form of a pulpy mass, into the pyloric caeca, where digestion is completed. Within three hours, the skeleton has been stripped of living tissue, and a few days later, algae have already established themselves there. As a result of the massive occurrence of Crown-of-thorns starfishes, more than 400 square kilometres of the Great Barrier Reef at a depth of 0 to 65 m are today stripped bare. Similar damage has been reported from the Truk Islands, from Saipan, Palau, Samoa, Hawaii, Guam and other areas of the Indo-Pacific.

The reasons for the sudden population explosion of Crown-of-thorns starfishes are not known for certain. Conjectures alternate between alterations in gene structure and harmful anthropogenic influences—radioactive fallout, pesticides, chemical pollutants and effluents. The following theory seems plausible. One Crown-of-thorns starfish produces millions of eggs, from which planktonic larvae develop. Normally, the majority are eaten by corals. In doing this, the corals effectively restrict the numbers of

33　Even the Crown-of-thorns starfish has its enemies. Here a 40 cm long Blowfish (Arothron hispidus) fearlessly snaps off the strong spines before biting into the starfish. (From a photograph by P. Vine in Ormond and Campbell, 1974)

their worst enemy. But if some of the extremely sensitive corals are killed by pollution of the water, there are suddenly millions of the more resistant larvae alive in the reef. As they drift to neighbouring reefs, they develop into minute starfishes. These begin to graze on the corals and as they grow, destroy them. The corals that typically consume the larvae are now destroyed by the Crown-of-thorns starfishes. The numbers of *Acanthaster planci* continue to grow, the plague spreads. A suggestion that the over-intensive fishing of snail shells is to blame for the disastrous proliferation of Crown-of-thorns starfishes is improbable. It is undoubtedly true that the Great Triton (*Charonia tritonis*) is an enemy of the starfish. But it is unlikely to attack a starfish more than once in six days, and in any case, the latter has a remarkable capacity for regeneration. The idea that population explosions of Crown-of-thorns starfishes normally occur at regular intervals is hardly tenable in view of the huge, living coral colonies that are several hundreds of years old, and which would never have reached such an age if Crown-of-thorns plagues were indeed periodic. So far, all attempts to bring the *Acanthaster planci* outbreaks under control have failed. Nevertheless, some of the reefs are returning to life

after decades, with a growth of young stony corals. Others even now are still colonized by algae, sponges and soft corals, and it may be decades before the former biological balance is restored.

Among the fishes of the reef, there are relatively many that feed on stony corals. Depending upon the manner of feeding, they can be divided into 3 types: those that pluck, those that gnaw and those that snap. Pluckers have small teeth set at the forward end of rather projecting jaws. With them, they successfully pluck coral polyps and coral tissue. Reef fishes of different families tend to be prognathous (Fig. 34). Most of them are Butterflyfishes of the genera *Chelmon, Forcipiger* and *Heniochus,* belonging to the family of "bristle-toothed" Chaetodontidae (Ill. 4, 85, 122, 131, 134). Filefishes (Monacanthidae) are represented by the small, green, yellow-spotted *Oxymonacanthus longirostris,* Doctor- or Surgeonfishes by the Moorish Idols *(Zanclus),* Rabbitfishes (Siganidae) by the Foxface *(Lo vulpinus)* and labroids by the Bird Wrasses *(Gomphosus).* Using their elongate snout, many of them also extract worms, crustaceans, snails and algae from their hiding places. This extreme prognathism can be interpreted as an adaptation to life in the coral reef. Thus fishes have taken over an ecological niche in the feeding network of the reef, making use of their "pincer" mouths to draw prey from small apertures. In the Caribbean reef region, only *Prognathodes aculeatus* occurs as a rather less well-developed representative of the pincer-mouth type of fish, while all the others live in the Indo-Pacific. But among fishes that pluck at stony corals, there are also some that do not have projecting jaws, such as the wrasses *Labrichthys unilineatus, Larabicus quadrilineatus* and *Diproctacanthus xanthurus.*

The gnawing feeders, Parrotfishes and Triggerfishes, have sharp, powerful teeth. With them, they gnaw at the live tissue of stony corals and at short algae lawns.

Their teeth marks are frequently to be seen on the surface of small-polyped, massive colonies as light-coloured, more or less broad scratches. The coral colonies are gnawed at superficially by the smaller fishes of this group, without suffering serious damage, indeed the skeletons, polyps and tissues are capable of regeneration within a short time. But the large species bite deep into the coral polyps. The damaged parts degenerate and are soon covered by a prolific growth of algae, sponges and other organisms. The coral colony goes on growing round about, producing torose margins in the course of years.

In addition, the strongest of the Parrotfishes snap their food. With their incisors shaped like a parrot's beak, they crack off the ends of branches noisily from shrub-like stony corals (Ill. 117), grind them in their powerful plate-like pharyngeal teeth and convey the calcareous pulp that is their food to the stomach. Since the proportion of digestible material contained is small, it is necessary for them to spend most of the day eating, and they constantly excrete faeces in the form of large clouds of chalky dust. About a third of bottom sediment in the reef is believed to have passed through their digestive tract. Other coral snappers include the Trunkfishes (Ostraciontidae). Up to 1 kg of coral, in quite large pieces, has been found in the stomach of large Porcupine fishes *(Diodon hystrix)* (Schleiden, 1888). Apart from environmental pollution, heavy seas caused by tropical storms, boring organisms and epidemics, fishes represent the principle disruptive factor for stony corals.

Reef corals secrete mucus, and do so in particularly large quantities when there is heavy wave action and a high level of sedimentation. The wax esters contained in it provide an additional source of energy, with 5.2 cal/mg. Mucus feeders include certain reef fishes and crabs such as *Trapezia.* Cyclopes also

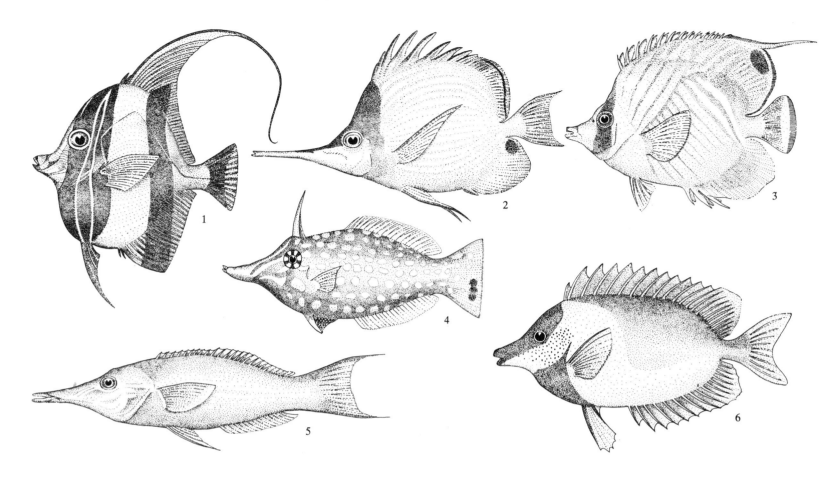

ingest mucus and utilize 50 per cent of its organic components as food.

Not only does mucus lie on the surface of corals but also floats in floccules or appears as a finely-dispersed cloud in the water. Accordingly, mucus-feeders are not always grazers but some of them make use of this food source as plankton-feeders and consumers of food particles. The latter take up fragments from the substrate. Many of them live on or in sand among the reef rocks. Irregular sea urchins, Acorn worms *(Balanoglossus)*, Polychaetes and crustaceans take in organic particles from it. The introvert and tentacles of Sipunculid worms (Sipunculidae) can be extruded by hydrostatic pressure, and are furnished with powerfully beating cilia. Particles settling on them are directed to the mouth. As the introvert is retracted, water is ejected in a spray from the mouth. Sea Cucumbers—*Holothuria,*

Halodeima and *Thelenota*—shovel loose calcareous sand into their mouth with the tentacles (Ill. 56). They fill the intestine 2 or 3 times within 24 hours, since the proportion of digestible particles—alga fragments, protozoa, larvae, small molluscs—is comparatively small.

Nocturnal Brittlestars fill the stomach three-quarters full of detritus, diatoms, foraminifers, radiolarians and planktonic organisms that have settled on the ground, all of which they pick up by dabbing at them with their numerous tube feet. Starfishes of the genera *Luidia,* *Linckia* and *Henricia* also seem to feed exclusively on small particles of matter. They sweep them along special channels to the mouth by means of currents created by cilia on the skin surface. The Brittlestar *Ophiocoma scolopendrina* has evolved a special way of taking in food particles. On tidal coasts, when humidity is low, the sun quickly dries up the

34 *The elongated jaws (prognathism) of many reef fishes belonging to different family groups are particularly suitable for plucking algae and coral polyps and for foraging in narrow crevices and holes for organic prey.*
1. Moorish idol, Zanclus cornutus *(Fam. Zanclidae),*
2. Long-Bill, Forcipiger longirostris *(Fam. Chaetodontidae), 3. Butterflyfish,* Chaetodon auriga
(Fam. Chaetodontidae), 4. Longnosed Filefish, Oxymonacanthus longirostris *(Fam. Monacanthidae),*
5. Birdfish, Gomphosus varius *(Fam. Labridae),*
6. Foxface, Lo vulpinus *(Fam. Siganidae) (Original)*

exposed lagoons and reef crowns, so that, as the tide rises again, detritus floats on the surface of the water. When the water has reached a depth of only 1–2 cm above the ground, the Brittlestar that is concealed among stones, stretches two of its arms up to the surface of the water, moves them towards one another like scissors, and transfers floating food particles that have been enclosed instantly in a mucilaginous cover, to the mouth. Within a few minutes, the water has risen to a depth of 5 cm, and the Brittlestars revert to their practice of picking up food particles from the ground. They also climb on corals to collect edible fragments. Many crabs strip the covering from plants with their claws or use their bristly gnathal appendages in a sweeping action. The nudibranch *Melibe bucephala* has a large round mouth with hair-lined edges, by means of which it brushes edible particles from algae. For sea urchins living enclosed in rock cavities, the quantity of detritus washed in through the narrow opening is sufficient to provide all the nourishment necessary for survival. The stomach contents of *Heterocentrotus mammillatus* has shown a predominance of foraminifers.

In coral reefs, as in other habitats, no predator exterminates the species on which it preys, and animals that are preyed upon are always available in sufficient quantities to ensure the continuation of the predatory species. Predators take vagile prey or parts of them. The animals likely to be caught are primarily sickly, weak or elderly individuals. As a result, the animals involved in reproduction are predominantly healthy, powerful ones that can produce a healthy progeny. A distinction can be drawn between major predators and minor predators. Minor predators take animals considerably smaller than themselves that belong to a lower systematic unit. The quarry of major predators (Ill. 114, 115), on the other hand, is found among animals of approximately the same size or larger,

and either related to them or higher on the systematic scale. Between the two groups, there are many transitions.

As far as is known, minor predators in the coral reef are mainly represented by crustaceans, echinoderms and fishes. Crabs seize ringed worms, bivalves and snails with their claws, crack them open and reduce them to small pieces (Ill. 116). Many brittlestars use the claw-like tips of their arms to grasp annelids, crustaceans, small snails and bivalves. Starfishes locate their prey by the perception of chemical and physical stimuli. Like the Crown-of-thorns starfishes, they envelop the victim in their everted stomach and pre-digest it outside the body. Many of them are able to apply suction with the tube-feet to force open the valves of mussels. Since the adductor muscles of the latter are very powerful, it may be hours before they yield and the valves part. But the eversible stomach can be insinuated into valves through an opening no more than 0.2 mm wide. Other species swallow their prey whole. Mussel shells, snail shells, remains of brittlestars, sea urchins, crustaceans, ringed worms, sea anemones and sea squirts have been found inside them.

At night, the Octopus glides across the rock. Crustaceans, mussels, snails and fishes are its principal food. Its sucker-bearing tentacles pry searchingly into crevices and cracks in the rocks. In a flash, prey are seized, drawn out, enclosed in the web of skin at the base of the arms and transferred to the mouth. A poison produced in the salivary glands kills the victim instantly. With sharp jaws like the beak of a parrot, it bites a hole in the shell or protective covering, injects gastric juices and later, sucks out the food pulp. It gnaws at mussels and opens them with its powerful arms. Even though octopuses may seem amusing creatures and divers may enjoy playing with them, because of their large-eyed, comical appearance and ability to alter the colour and character of their body

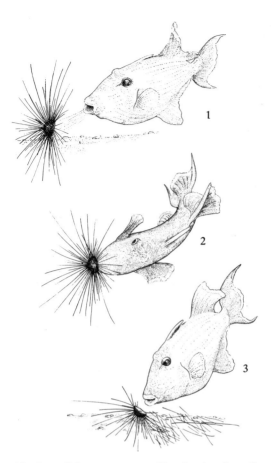

35 Some fishes are even capable of eating the well-armed Long-spined urchin. The Triggerfish, Balistes fuscus, overturns the sea urchin by directing powerful jets of water at it (1), bites the short spines from the underside (2) and cracks the limestone shell (3). (After Fricke, 1972)

surface depending upon their state of stimulation and the immediate background, or to make themselves invisible in a cloud of "ink", caution is advised, particularly in handling the smaller species. Their poison is so toxic that a bite can be fatal to a human victim. Large species are rather rare in the coral reefs, and the related Giant Squids that can be as large as 20–30 m, with eyes the size of soup plates, live rather in the neighbouring deep-sea regions.

White Tip reef sharks *(Triaenodon obesus)*, Nurse sharks *(Ginglymostoma)* and many other shark-like fishes as well as large Stingrays (Dasyatidae) rove about the reefs and over sandy and stony ground, searching for crustaceans, molluscs and echinoderms (Ill. 126). They also track down prey concealed in crevices in rocks, among scree and in the sand. Using sensory organs that respond to electrical impulses, they are able to perceive and locate ciliary currents set up by the prey. These electro-receptors are the ampullae of Lorenzini, situated in the region of the snout. Dragonets *(Callionymus)* and many other osseous fishes catch ringed worms and small crustaceans.

Even the Long-spined Hatpin urchins have enemies. Triggerfishes eject powerful streams of water from the mouth and wash prey out of the sand (Fig. 35). *Balistes fuscus* even uses this method to force Hatpin urchins from their refuges in the rocks. It directs streams of water at sea urchins lying on open ground to overturn them, hurls itself at the mouth side where the spines are short, cracks the test with powerful teeth and consumes the contents. If sea urchins have wedged themselves very securely into a rock crevice, it gradually bites off the spines, apparently little perturbed if they pierce its head. Finally it eats the urchin body from above. Certain Snappers *(Lutjanus)* consume sea urchins whole, spines and all. The large wrasse, *Cheilinus trilobatus*, lifts sea urchins up by the spines, grasps them on the underside

with its mouth as they float in the water and smashes them against stones. The White-spotted Blowfish, *Arothron hispidus*, and the Triggerfishes *Balistoides viridescens* and *Pseudobalistes flavimarginatus* are prepared to eat even the Crown-of-thorns starfish (Fig. 33), from which they snap the poisonous spines with their powerful teeth and then bite the calcareous shell to pieces. The huge Humphead Wrasse *(Cheilinus undulatus)* catches snails, fishes, sea urchins, bivalves, crustaceans and brittlestars by ramming its back violently against branching corals and snapping up the startled creatures that emerge from it. It also consumes species with sharp or poisonous spines such as the Crown-of-thorns starfish.

Major predators are found predominantly among flatworms that are often brilliantly coloured, inconspicuous threadworms, ribbonworms and ringed worms, snails, cephalopods, crustaceans and fishes. Ribbonworms (Nemertini) that are up to 70 cm long, mostly annular in cross-section, and usually very thin, suck out the bodies of snails and worms. The larger ones swallow molluscs, ringed worms, small crustaceans and fishes. They pursue their victim with rapid, undulating, crawling movements, encircle them with their long proboscis and draw them into their mouth. Small, predatory threadworms (Nematodes) take rotifers, annelids or similar creatures. They either swallow them whole, tear pieces from the bodies with their vicious teeth or suck them dry.

A surprisingly large number of major predators are to be found among snails. Certain predatory snails swallow large prey whole. A Great Triton or Sea Trumpet 28 cm long swallows a 21 cm long Sea Cucumber in 4 hours. Small Moon snails (Naticidae) drill a hole with their radula in the shells of other snails, extend their proboscis into it and eat their quarry. A number of Rock shells (Muricidae) also drill holes in this way.

129 94

94 *Only as a larva is this crustacean fully mobile. When it has attached itself to a base, the Goose Barnacle* (Lepadomorpha) *produces a stalk, shell-like plates and protruding feathery tentacles with which it catches plankton suspended in the water. (Red Sea)*

130/131 95 96 | 101
97 98
99 100

95 *A yellow sabellid has found a favourable site for the passive capture of plankton on a raised Micropore coral* (Montipora meandrina). *(Red Sea)*

96 *The Fan or Feather-duster worm,* Sabellastarte indica, *has settled "head downwards" among the disc-shaped colonies of Undulate coral* (Leptoseris incrustans). *(Society Islands, Pacific)*

97 *Some hydrozoa form impressive colonies of a chitinous substance, which add to the extreme diversity in types of growth among sessile organisms living in a reef. (Red Sea)*

98 *The extended polyps of this Sea Mat* (Zoantharia) *are reminiscent of small, fringed Parasol mushrooms. (Red Sea)*

99 Boloceroides mcmurrichi, *a sea anemone only 1 to 3 cm in size, frequently detaches itself from its base, pumps itself full of water and swims along with rhythmic strokes of its tentacles. (Red Sea)*

100 *The Porcelain Cowrie* (Cypraea) *glides across the surface on its broad foot with its mouth protruding like a proboscis. The significance of the dense fringing of the mantle is not known. (Red Sea)*

101 *The two rings of tentacles of the Feather-duster worm,* Spirobranchus giganteus, *are limbs that have been modified to become an effective apparatus for catching plankton. (Red Sea)*

132 102
103 104

102 *The Worm Cucumber (*Synapta sp.) *is a typical creature of the lagoon lying behind the outer reef. With its fringed tentacles, it takes up vegetable matter from the sand. (Red Sea)*

103 *A brownish-yellow sponge has grown into a large colony of "chimneys"; water drawn in through pores on the sides is expelled through the oscula or upper openings. (Red Sea)*

104 *The red sponge* Axinella *is very common at depths of 15 to 25 m or in caves. (Red Sea)*

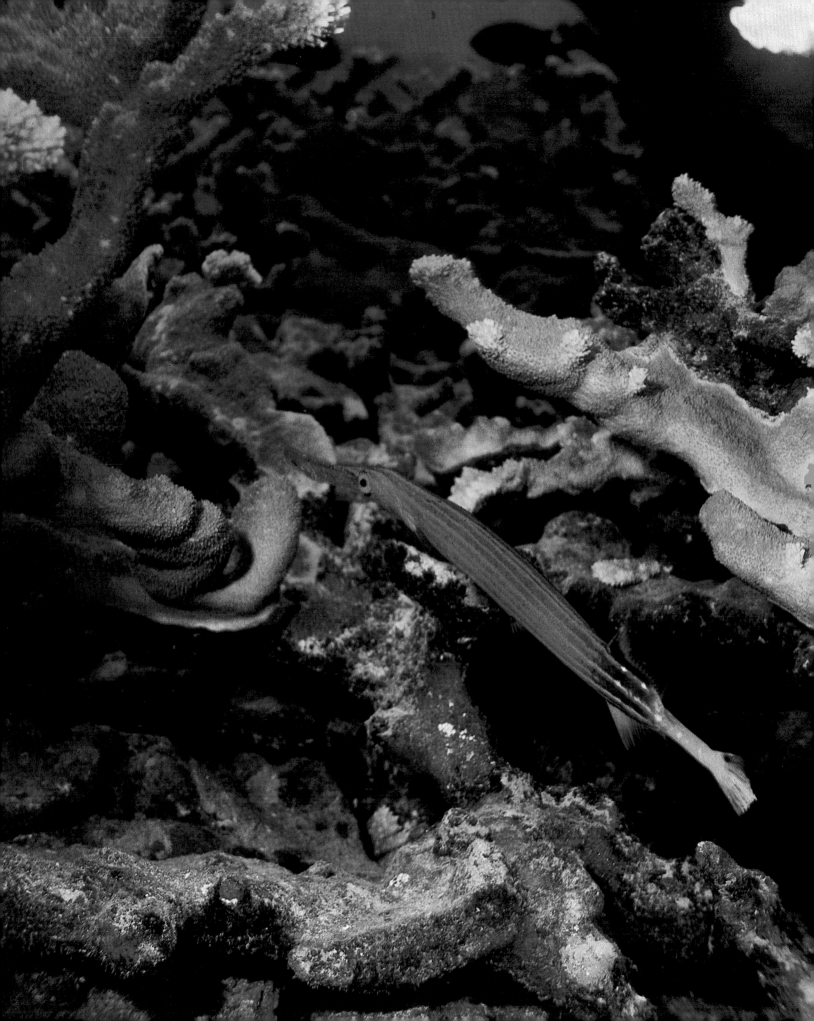

Snail shells in which a hole has been drilled can often be found on beaches. Other Rock shells force a tooth into the narrow slit between the valves of a mussel, prizing them apart. The murex then inserts its proboscis that is furnished with rasping teeth, and eats the creature. Helmet snails (Cassidae) have a foot covered in very thick skin which secretes large quantities of mucoid slime lubricant. Protected in this way, they can climb onto sea urchins without being hurt by spines, and paralyze them with a poison. They then secrete an acid that renders the skeleton brittle so that it can be pierced by the radula. Vase snails (Vasidae) crawl on sand. With delicate sensory organs, they detect snails and ringed worms living in the ground. They insert the proboscis and catch the victim. Olives (Olividae) burrow through sandy ground and extend their long breathing tube into the open water. If they pick up the scent of prey, they burrow towards it underground and unobserved. They emerge so suddenly in its immediate vicinity that they are able to catch the worm, crustacean or fish in the forepart of the foot before it is able to take flight. As they push the prey towards the mouth, they burrow into the ground.

The most highly specialized predators are the Cones *(Conus)*. Each of the more than 500 species pursues a different prey and in this way there is no competition for food. They feed on molluscs, ringed worms and fishes. The radular teeth have evolved into long, hollow cannulae with barbed and pointed tips, which are connected to poison glands. The grooved dart is always ready for action at the end of the proboscis. Reserve teeth are held in the radular sac to be drawn out for use one by one. To catch prey, the proboscis is extended, the cannular tooth thrust out to pierce the victim and a powerful toxin injected. The prey is swallowed when it is paralyzed or dead (Fig. 36). The poison is so strong that its effect on humans can be fatal.

In reefs and lagoons, Sepias or cuttlefishes and Squids have been observed drawn up in single file or in line abreast, as if in military drill formation. They swim rapidly in pursuit of fishes and seize them with their tentacles. The Squid, *Loligo,* bites the head from a fish and eats the body, apart from the gut, from the front end to the back.

Few crustaceans are major predators. Mantis shrimps (*Squilla* and related species) have sharply barbed raptorial legs that they hold folded in like closed pocket-knives, rather in the manner of the terrestrial Preying Mantis. They inhabit scree-like ground and seagrass meadows. At twilight, hidden motionless in the darkness of tubes in which they live, they keep watch with their stalked eyes for suitable prey. If a small fish or crustacean approaches, a leg strikes out with lightning speed. There is no escape. Pistol shrimps *(Alpheus)* are some 2–4 cm long with a greatly enlarged claw (cheliped). The immovable finger of the latter has a groove terminating basally in a cavity. The movable "thumb" (dactylopodite) is equipped with a process that fits exactly into the cavity. If prey or an enemy approaches, the movable thumb is brought down with such force against the immovable finger that the water, forced suddenly out of the cavity, is driven with a loud report along the channel that functions as a rifle barrel, directing a powerful stream of water against the prey. The latter is certainly surprised and may even be stunned, and the small shrimp is able to seize it immediately and drag it into its refuge.

In addition to speed and a powerful bite, the main ingredient of success for many major predators in the fish world is patience. Groupers remain invisible in dark caves and gorges, lying in wait for prey (Ill. 114, 129). With their bulging eyes rolling, they follow every change in the surroundings. They seize their victim when it is quite close. Other predators lie freely on open ground and are nevertheless invisible. Coloration, markings, body shape and skin appendages make them so similar in appearance to rock or algae that they merge into their background. The Stonefish *(Synanceja)*, Scorpionfish (*Scorpaena* and related species), Flathead *(Platycephalus)*, Frogfish (Antennariidae) and Carpet Shark *(Orectolobus)* are all masters of camouflage (Ill. 142). When prey comes close to them, they simply open wide their jaws. The suction that is produced sweeps the prey into their mouth. Red Firefishes *(Pterois)*, also variously known as Turkeyfishes, Dragon- and Lionfishes, move in the darkness of caves, low down close to the ground (Ill. 135, 137), with their great divided fins spread wide, and drive their prey along in front of them and into a corner. Here, they snap them up. Both the camouflage adepts and the pursuers eat fishes and crustaceans. The great twilight hunters are lively movers. During the minutes of sunrise and sunset, Reef sharks, Barracudas and Sea basses attack the diurnal fishes that are still drowsy, tear them up or swallow them whole. Morays are also on the move at night. An acute sense of smell leads them to their prey.

Among fishes, there are certain "wolves in sheep's clothing" — predatory or they could be called parasitic species, in the guise of harmless ones. For example, small Blennies (Blenniidae) of the genera *Aspidontus* and *Runula* resemble Cleaner fishes (see p. 144) in shape and colour (Fig. 42, 2 and 3). Since they also move in a similar way, other fishes swim to them to have themselves cleaned of parasites. This is when the imposter attacks, rapidly tearing fin tissue, scales and flesh from the victim's body with its sabre-like teeth.

Although probably the most highly predatory of all fishes, they measure no more than 7–12 cm, and so the large fish that they have hoodwinked, escapes with only a fright and a few not very dangerous bites.

Dead animals never lie for long in the reef for scavengers consume them. Apart from bacteria and microscopically small organisms, fishes, starfishes, crabs, crayfishes and a large number of snails eat putrescent animal matter. Rock shells force their proboscis into the opening of the shell of dead crustaceans and eat the insides. Predators also eat carrion that is not too highly putrid, for example, Tritons consume the dead bodies of cephalopods and fishes.

Large numbers of parasites live among the diverse organisms of the reef, although so far, little is known about them. Ectoparasites suck juices from host animals or consume fragments of their tissues. The isopod, *Allophryxus malindiae,* that was discovered only a few years ago, is parasitic on the shrimp *Coralliocaris superba* that lives on corals. It is not unusual for isopods to attach themselves firmly to the head or branchial arches of fishes. Species related to cyclops are sometimes embedded into the gill lamellae of fishes. They have reduced the limbs and taken on a saccular or tubular form, adapting in this way to a parasitic way of life. Parasitic snails (Pyramidellidae) attach their proboscis firmly to worms, coral and polyp organisms, pierce them with a fine, hollow stiletto and suck out the body fluid. Others take in host fluids through the skin. The snail *Pelseneeria,* 4–5 mm in length, lives among the spines of sea urchins and sucks food from them. Apparently the urchin's pedicellariae are unable to affect the parasites that are protected by a thick covering of mucus. Endoparasites such as amoeba, flagellates and infusoria, tapeworms and roundworms live primarily in the digestive tract of annelids, crustaceans, molluscs, echinoderms and fishes, as well as in other organs and within muscle tissue.

A graphic example of the wide variety of parasites is provided by Sea Cucumbers, in their role as host organisms; because of their lack of defences and

36 *Using its chemical sense organs, a cone, having located a Goby that is guarding its spawn and therefore will not move from the spot (1), propels a minute poison dart (shown enlarged in the sketch) through the proboscidean extension of the pharynx into the body of the victim (2), paralyzing the small fish which it then swallows (3). (From a series of photographs by A. Kerstitch, 1980)*

their immobility, they are particularly heavily affected. Small ringed worms live among the many excrescences and papillae of the body covering. Within the digestive tract and body cavity are protozoa. A cyclops may be found in the oesophagus. Flatworms, mussels *(Entovolva)* and the worm-like, blood-feeding snail *Entoconcha mirabilis* live in the intestinal canal. Crabs of the genus *Pinnixa* feed on excrement in the cloacal region. Even certain fishes, the Cucumber fishes (Carapidae), also known as "Pearlfishes", have adapted to a parasitic life in the spacious branchial gills of the Sea Cucumber. They feed on the reproductive organs, lungs and other internal structures and are very well protected. Juvenile stages of these fishes are unable to live for long outside the host. They have been found in the starfish *Culcita novaeguineae* as well as in Sea Cucumbers. The adult fishes, that are able to maintain themselves in open water for a longer period, also prey on microorganisms. When threatened, they immediately retreat headfirst or tailfirst through the anal aperture into the cucumber (Fig. 37). The species *Carapus homei,* that is less highly host-specific, lives in a similar way in pearl oysters. Small relations of jellyfishes of the species *Tetragonurus* can occasionally be found inside planktonic Sea Cucumbers *(Salpa, Pyrosoma)* that drift sporadically into reef regions at night. The parasites are protected inside the transparent tunicates and also feed on them, cutting fragments from the host tissue with their curious dentition that resembles a saw.

Among the parasites on stony corals, unicellular organisms and vermiforms have not so far been extensively researched. Snails belonging to the families Epitoniidae and Coralliophilidae— *Coralliobia, Leptoconchus, Magilus* and *Coralliophila* —are always associated with particular corals and also feed on them. Fossil evidence shows that their ancestors still had normally coiled shells and lived outside on the coral colonies. The forms of recent species exhibit progressive morphological adaptation to the parasitic habit. For example, in the presence of *Quoyula madreporarum,* Styloid corals develop a shallow lamelloid basal structure into which the rim of the snail's shell is fitted with great precision. The ovate *Leptoconcha* and *Coralliobia* drill holes in coral skeletons in which to live. With their long, extensible proboscis, they suck food from polyps in the vicinity. The juvenile form of *Magilus antiquus* that lives in stony corals, still has a coiled shell, which in the course of further growth, develops into an irregular tube. *Coralliophila* lives externally. Its salivary secretions penetrate the epidermis of the coral tissue, which is then digested and sucked out. The mussel *Fungiacava elatensis* that is the size of a grain of corn, and which allows the skeleton of fungus corals to grow round it, extends its siphon into the gastrovascular cavity of the host and extracts water, oxygen and food. The bodies of parasitic cyclopes that are on stony corals, as on many other organisms, show more or less extensive deformation, depending upon the degree of adaptation (Fig. 31).

The great abundance of food and the competitive pressure in the coral reefs have led to many examples of specialization in various groups of animals in the course of their evolution and in the continuous process of speciation. For example, food-getting habits of half of the approximately 90 species that make up the colourful *Chaetodon* genus of Butterflyfishes are fairly well-known. 13 species feed on a wide variety of benthonic organisms such as worms, crustaceans, small snails, coral polyps, fish eggs and algae, two species on algae and plankton, one species exclusively on algae, two species exclusively on plankton, four species on algae and coral polyps, 17 species pluck coral polyps either predominantly or exclusively, and one species even functions as a cleaner.

Probably the omnivorous species are the original form. From them, specialists have developed in the course of evolution. Obviously butterflyfishes are still subject to a dynamic radiation process, since a series of examples of hybridism can be observed in the reef—a criterion of the high plasticity of genetic factors. Analogous tendencies can also be observed among other animal groups rich in species in coral reefs.

Interaction between reef inhabitants

In coral reefs, myriads of organisms living in close association share living space and food. They protect and defend themselves, clean themselves and procreate. Only by the efficient functioning of particular mechanisms of communication—by mere presence, coloration, stance, posture and the articulation of sounds—is rapprochement possible, distance maintained or retreat initiated. Reactions are based both on inter- and intraspecific relationships. Each individual respects each other one, and the result of this attitude of "not too much and not too little", of not remaining in

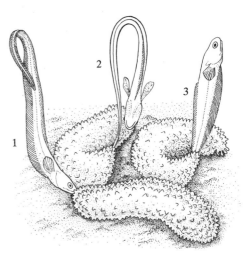

37 A Cucumber fish, Carapus sp., wanting to enter a Sea Gherkin, causes the host first to extend the vent (1), introduces initially the tip of its tail (2) and then the whole body (3). (After Norman, 1966)

too close proximity, of not producing too many progeny—the result, in short, is that by maintaining the biological balance, the survival of the species is ensured. A few examples will illustrate the rules that guide the interaction between reef organisms.

Sessile animals such as soft corals, gorgonians and thorny corals, sea anemones, bryozoa and bivalves are inhabited by a remarkable diversity of other organisms and used as a feeding place and refuge. From the brilliantly pigmented mantle margins of Giant Tridacnas, specialist crustaceans gather particles of food. Above the bivalve's incurrent siphon, fishes swim and snap up planktonic organisms that are whirled along in the inflowing water. Molluscs, vermiforms and minute crustaceans as well as a vast number of microscopically small protozoa live in sponges with their many pores and chambers. Smaller species of fishes are permanently or occasionally dependent upon sponges. Young labroids of the species *Clepticus parrai* spend the night head-downwards in the inner cavity (spongocoel) of sponges. Certain fishes spawn in sponges; the eggs are well protected and are constantly supplied with fresh oxygen-rich water by currents set up by the sponge. *Trypton* shrimps live in a group of several at a time in a sponge. They ingest particles from its surface as food and in this way, keep it clean. But occasionally they bite fragments from the sponge tissue, and so live sometimes as parasites, sometimes as symbionts.

Many vagile animals also have cohabitants or associates. By means of stalked multi-valved mobile pedicellariae in association with poison glands, sea urchins prevent their surface plates from being colonized by clinging microorganisms, but the ends of the long spines that cannot be reached are immediately colonized by epizoa. Crawling Comb Jellies (*Coeloplana*) that seek protection within the covering of spines, small sea gher-kins, brittlestars, ringed worms, snails, cyclopes, ostracodons, sea fleas, shrimps and crabs are tolerated, although some of them, as true parasites, damage skin tissue. Cardinalfishes of the genus *Siphamia* hover above long-stalked sea urchins snapping at plankton, and immediately when an enemy approaches, disappear into the dense forest of spines (Fig. 38). *Siphamia fuscolineata* rests during the day among the spines of the Crown-of-thorns starfish. Other echinoderms—star fishes, basket stars, feather stars, sea cucumbers—also live in close association with organisms of alien species. Most of them are harmless cohabitants, some of them slightly damaging, a few beneficial as cleaners.

Fast-moving, agile crustaceans and fishes establish a territory that they defend relentlessly against intruders. Damselfishes mark the limits of their territory by threat displays. They swim at intruders with raised fins, produce snarling noises by rubbing together the gill covers, or bite at the interloper. The size of the enemy seems to be irrelevant. *Pomacentrus flavicaudus* defends an area of some 2 m² against Damselfishes, Parrotfishes, Surgeonfishes, Rabbitfishes, Butterflyfishes, Puffers, Blennies, Gobies and Wrasses. Damselfishes living in a group within a territory, defend it in common.

Recognition of an enemy is always selective, and is coupled with a differentiated defensive instinct. It is linked to factors such as competition for food and brood care, and varies from species to species. *Eupomacentrus planifrons* attacks intruders at different distances. It clearly looks upon some members of its own species as its principal rivals and swims at them as soon as they approach within 4 m. It allows members of the same genus to approach rather more closely, but differentiates between them. Of *Eupomacentrus fuscus, E. variabilia* and *E. partitus,* it permits the latter to come closest to it. Of non-related species, the Rock Beauty *(Holocanthus tricolor)* is attacked at the greatest distance, namely 2.5 m, but the small Harlequin bass *(Serranus tigrinus)* not until it is quite close. The same Damselfish, that feeds on algae, even removes sea urchins from its grazing grounds as competitors for food. It bites the spines from Long-spined Hatpin urchins, alarming the creatures so that they move away. It even carries the short-spined *Echinometra* and *Lytechinus* out of its territory.

In 1976, Fricke observed Damselfishes of the genus *Dascyllus* that inhabit branching corals in groups (Ill. 87), and have developed a prophylactic method of defense, which, by analogy with the chasing of birds of prey by small birds, can be termed "mobbing". A Moray that had caught a Three-spot Humbug Damselfish *(Dascyllus trimaculatus)* and was making off with the victim in its jaws, was harried and bitten by the members of the group. Such behaviour might well encourage the Moray in future to avoid branching corals together with the fishes that inhabit them. An octopus that attempted to settle inside the territory was persistently harassed and bitten by the nimble Damselfishes until it moved off again.

Sexually mature Striped Parrotfishes, *Scarus croicensis,* defend their territory of approximately 12 m² in pairs. The

38   *When danger threatens, Cardinal fishes,* Siphamia versicolor, *seek protection within the thicket of a sea urchin's spines. (After Eibl-Eibesfeldt, 1967)*

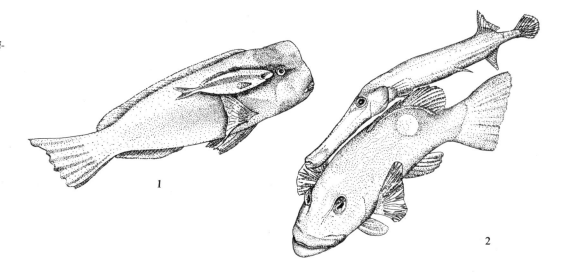

39 Jacks (1) and Trumpetfishes (2) swim as "riders" above and in close alignment with large fishes and dart forward from this camouflage position to capture prey. (From photographs by Kühlmann (1) and Fricke (2))

1

2

boundaries are marked by mouth-to-mouth fighting with members of the same species. Sometimes a second male with one to three subordinate females is tolerated. These females do not engage in territorial defense. Demarcation of territory is also carried out on a small scale. For example, the Goby, *Paragobiodon echinocephalus,* that lives on the Styloid coral *Pocillopora damicornis* in groups of several individuals, indicates its possession of territory to the crab *Trapezia cymodoce* and the Pistol shrimp *Alpheus lottini* by vibration of the body. The crabs understand the gesture and keep their distance. The shrimp remains close to *Paragobiodon,* its antennae in contact with the tiny fish. In this way, any agitation that the fish experiences in face of a threat is immediately transmitted to the shrimp. The allegiance of a fish to a territory is illustrated by experiments in marking carried out on the wrasse *Tautogolabrus adspersus.* Even after being kept for nine months in an aquarium, the fishes found their way back to the original home from a distance of 4 km.

Whereas territorial behaviour keeps animals apart, the search for food often brings different species together. It is not unusual for one animal to live on the abundant food supply of another, even more commonly on their leavings, without causing them any harm. They

are known as commensals or table companions. Fishes that stir up sand or dig into the substrate in search of food are often accompanied by other fishes. For example, swift and nimble wrasses of the species *Halochoëres centriquadrus* follow schools of Red Mullets (Mullidae) as they move across sandy ground energetically rooting for food. Any animal matter that the mullets turn up but do not take themselves is snatched away by the more quickly-moving labroids. In addition, they resemble the red mullets in their markings, so are less conspicuous as food competitors. Wrasses of the species *Thalassoma rueppeli* arrive quickly on the scene when the dark-coloured Triggerfish *(Balistes fuscus)* consumes a sea urchin. In the same way, other fishes swim round Rays as they dig in the ground for bivalves, snails or worms. Off the southern coast of Cuba, a large Stingray *(Dasyatis americanus)* has been observed, followed by half a dozen sizeable Jacks *(Caranx)* that pounced on food organisms in the clouds of sand thrown up by the rays. Wrasses swarm round divers as they break off pieces of reef rock and corals and in doing so, disturb numbers of worms, molluscs and crustaceans that are eaten in a flash by the fishes. They also follow large Parrotfishes that snap pieces from coral colonies with their strong beak-like teeth

and in this way expose small animals.

Interspecific hunting associations are particularly interesting. For instance, the Gomphosus wrasse or Birdfish, *Gomphosus caeruleus,* and the Goatfish or Surmullet, *Parupenëus chryseredos* sometimes swim through the reef together. The Goatfish uses the tactile barbels under its chin to detect small food organisms in cracks and holes, but is unable to reach them with its blunt snout with the mouth on the underside. Now the wrasse moves into action, draws the prey from its hiding place with its extended, beaklike mouth, and both of them consume it. The Groupers *Cephalopholis argus* (the Blue-spotted Argus or Peacock Rock Cod), *Aethaloperca rogas* and *Plectropomus maculatus* (the Striped Madagascan Grouper) (Ill. 119) also seem to hunt in pairs from time to time. While one swims into a rock cleft or cave, the other takes up its position outside the exit in order to catch the fleeing prey. *Cephalopholis argus* also lies in wait outside caves into which an octopus or moray has penetrated.

It is curious to see the Trumpetfish *(Aulostomus maculatus)* "riding" a larger fish, often a Parrotfish (Fig. 39, 2). The Trumpetfish swims above its "steed" in very close proximity, following its movements and changes in direction and depth exactly. If the pair come upon

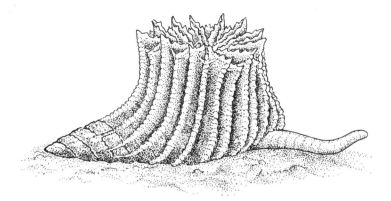

40 Well-protected from its enemies, the sipunculid worm drags migratory coral across soft ground. In this way, the very small coral avoids being covered by sand even when siltation is high. (Original)

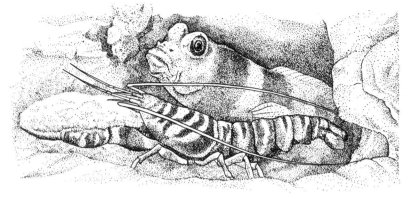

41 The blind pistol shrimp, Alpheus djiboutensis, uses its two pincer-legs held close together as a shovel on which it carries sand from the cave in which it lives with the watchful goby, Cryptocentrus lutheri. It maintains contact with the goby by means of its antennae and in this way picks up information from the behaviour of the fish. (From a photograph by W. Luther from Eibl-Eibesfeldt, 1967)

prey, the Trumpetfish darts forward onto the victim quick as an arrow from the cover of the harmless Parrotfish. The small Jacks, *Carangoides bajad* and *C. fulvoguttatus* align their body with that of large Humphead Wrasse, *Cheilinus undulatus,* or *Bolbometopon muricatus* as they hunt for prey (Fig. 39, 1). The small Sea Bass *Diploprion drachi* has even adapted to the twilight activity of the large Blue-spotted Argus, *Cephalopholis argus,* that it uses as a means of camouflage. Various small labroids that can modify their colour rapidly to that of their "mount" have also been observed in the role of "rider".

The practice of shoal formation is a protective adaptation. Its significance lies in the fact that in order to seize prey, predators have to take a bearing on a single individual. A cheetah, for instance, has to separate an antilope from the herd in order to strike at it. A barracuda following a shoal of sardines is confused by the myriads of flashing forms darting through the water. But it has a chance if it can cut off the retreat of an individual separated from the shoal. Only then can it take aim and strike. Fishes are better protected even in small groups than

alone. Many young predatory fish that later live singly, join together in "schools" as juveniles. Even young barracudas roam the waters in troupes. The larger the shoal, the poorer the chances of success for predators. Plankton-feeders that move along coasts, such as Sardines *(Sardinella)* and Silversides (Atherinidae), and which sometimes appear in reef areas, can survive only in large permanent shoals. Even quite small organisms of fixed habitat, such as Mysid shrimps, Cardinalfishes (Apogonidae) and Sweepers *(Pempheris),* spend the day in shoals in the shade of overhanging rocks and in dark caves. Fusilierfishes *(Caesio)* are dependent upon the reef because they sleep there at night. In the morning, they assemble in a shoal and set off in a body across the outer slope in search of plankton. In this, they exhibit behaviour convergent with that of Sand Lances (Ammodytidae) of the northern seas. Species of Grunts *(Haemulon)* form mixed shoals. Algafeeding Surgeonfishes and Parrotfishes relax strict shoal formation somewhat while grazing, but only to the extent that at any moment they can take flight as a body (Ill. 108). Fishes attached to partic-

ular territories and living in groups derive a further benefit from their knowledge of the locality. When an enemy appears, Jewelfishes *(Anthias)* and Damselfishes of the genera *Chromis* and *Dascyllus* disappear into particular openings in the rock or coral that are recognized as belonging to each individual by all members of the shoal (Ill. 87, 110). Fishes of the same species but alien to the group are driven away. A recognizable hierarchy exists in the group. In contrast to members of large migratory shoals, they have achieved a high degree of organization.

The remarkable phenomenon of symbiosis implies a mutually beneficial partnership between two individuals of different species. It is relatively common in the coral reef, because the pressure of competition resulting from the wealth of organisms existing there has continually brought about new adaptations, and symbiosis, as an extremely successful and advanced way of life, has come into being more frequently here than in other biotopes. Indeed, endosymbiosis between unicellular algae and stony corals is the very basis of the development of coral reefs (see p. 24).

Even more striking, however, are examples of ectosymbiosis that the diver can observe directly. Shrimps in particular often engage in symbiotic association with other animals (Ill. 146). Among the long, dark spines of the Hatpin urchin, the equally darkly coloured shrimp *Stegopontia* finds a hiding place. It even imitates spines ringed with bands of lighter colouring that sometimes occur. It takes up edible particles from the body of the sea urchin and so cleans it. The exchange of protection for body grooming also exists between sea anemones and shrimps. For instance, shrimps of the genus *Periclimenes* live within the protection of stinging anemone tentacles. When danger threatens, the shrimps, that have completely translucent bodies apart from a few white stripes and coloured spots, settle with limbs extended and antennae folded back so closely against the tentacles of the host, that they are quite indistinguishable from their background. They also remove fragments of dirt from the host.

Anemonefishes *(Amphiprion)*, also known as Clownfishes on account of their marking of whitish rings and masks (Ill. 125), are completely dependent upon actinians. They live in a sea anemone in groups of several at a time, and by nestling movements, cause it to extend its tentacles. At night, they even sleep inside the host's digestive cavity. Not only do they clean particles of dirt from the anemone, but they also drive the latter's enemies away by swimming at them, by

emitting warning "tok-tok" sounds and by aggressive displays of their brightly-patterned body. A transverse line of colour across the eye, which can be larger or smaller depending upon the inclination to attack, functions as an optical warning signal. Since fishes that penetrate the territory heed and interpret every sign, however small, there is rarely any biting; the intruders make off as soon as they become aware of the eye stripe. Anemonefishes distinguish between their enemies according to the degree of damage they are able to inflict. Their worst enemy is a *Lethrinus* Snapper, which they attack fearlessly even when it is still at a considerable distance from them. The Butterflyfishes *Chaetodon fasciatus* and *Ch. lunula* (Ill. 122), that feed on the tentacles of actinians, make a wide detour round sea anemones that are inhabited by Clownfishes, since they clearly fear the attack that would otherwise immediately result. The nimble labroids *Thalassoma* and *Halichoëres,* that seek food in every corner, are inveterate predators of spawn. So the large red clutch of eggs that the Anemonefish attaches to a rock under the protection of a sea anemone is at considerable risk. Anemonefishes launch a violent defensive attack on labroids even when they have no spawn to defend. In this way, they discourage the egg thieves from approaching anemones at any time. Sometimes Anemonefishes adopt a measure of parasitism, for, in addition to organic plankton and algae, fragments of tentacles belonging to the

hosts have been found in their stomachs. If Anemonefishes are driven by force from their living fortresses, they fall victim to predatory Rock Cods and Groupers, Lizardfishes *(Synodus),* Scorpionfishes *(Scorpaena)* or Jacks (Carangidae). How is it possible for fishes and crustaceans to live in symbiotic association with stinging anemones, while other creatures are driven off or paralyzed and killed by the poison? Even shrimps and Clownfishes are stung at first. But as a result of repeated contact, they progressively accumulate mucus from the anemones. The symbiont, now "smelling of" anemone, is no longer attacked.

On sandy ground among reefs, it is not unusual to find depressions that are the entrances to tubes. They are inhabited by highly specialized Pistol shrimps *(Alpheus)* and Gobies *(Amblyeleotris, Cryptocentrus* and *Lotilia)* living in pairs (Fig. 41). While the Gobies lie quietly in the entrance funnel, keeping watch for prey, the crustaceans, working incessantly like small bulldozers, shovel sand and small pebbles to the outside, using the enlarged pincer-leg bent inwards. As they do so, they extract food particles for themselves. They are rather like blinded slaves, for as cave dwellers, they have lost their sight. For this reason, they maintain constant contact with the Goby by means of one of their long antennae. When an enemy approaches, the Goby disappears immediately into the tube. Heeding the warning, the shrimp follows rapidly. Each affords effective protection to the other, the shrimp to the fish by

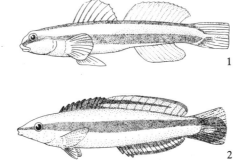
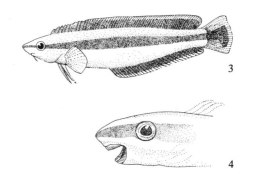

42 *All obligatory cleaner fishes, such as the neon goby,* Elacatinus oceanops *(1) and the cleaner wrasse,* Labroides dimidiatus *(2), are small, elongate and striped, as is the mimic* Aspidontus taeniatus *(3) which bears a strong resemblance to them, but which attacks fishes thus misled with its sharp saw-like teeth (4). (After Eibl-Eibesfeldt, 1967; and Matthes, 1978)*

building the tube in which they live, the Goby to the blind shrimp by providing an alarm system.

A highly developed form of association between different species for their mutual benefit is the cleaning symbiosis. Shrimps and fishes of various families, originally grazers, particle eaters or minor predators have, in the course of evolution, developed as cleaners of larger animals. They remove irritants, parasites, fungoid growths and skin impurities from their bodies. The acquisition of this capacity is to be seen as an adaptation, as a changeover from highly competitive feeding to a reliable food resource that had so far remained untapped. Cleaners have been observed at work primarily on fishes, but also on crocodiles and even on divers. They are particularly common in coral reefs.

A number of nocturnal shrimps clean ectoparasites and sloughed skin fragments from sleeping fishes. Diurnal species of *Periclimenes* that seek protection in actinians or *Stenopus hispidus* and *Hippolismata grabhami* that live in rocky lairs, signal by waving their long antennae (Ill. 147). Fishes in need of cleaning indicate their readiness by adopting a particular body stance. Only then does the shrimp leave its safe refuge and move across to the fish. During the process of cleaning, it maintains constant contact with the fish by means of its antennae. Groupers and Snappers willingly open their mouths to cleaners. The shrimp enters fearlessly and cleans the giant's teeth of food remnants. If the procedure lasts too long, the latter shakes its head. The shrimp rapidly emerges from its jaws and returns to its station to await the next "customer". The cleaner shrimp *Leandrites cyrtirhynchus* swims in open water in front of its hiding place in the rocks, in this way demonstrating its readiness to act as a cleaner. Its white markings make it conspicuous and fishes come to have their parasites removed.

Cleaner fishes behave in a similar way. The obligatory cleaners—Cleaner labroids *(Labroides)* and Neon Gobies *(Elacatinus)*—have a thick, dark longitudinal stripe on their flanks (Fig. 42). When ready to clean, *Labroides dimidiatus* swims with bobbing movements that might perhaps be a ritualized combination of swimming at an enemy and taking flight. By means of this "cleaning dance", it draws attention to itself and at the same time, elicits in the larger fish a readiness to be cleaned and an inhibition of the urge to feed, that is expressed in a variety of ways by different species. Perch-like fishes freeze, open their jaws and extend the gill covers stiffly. Others align themselves vertically, lie on one side, draw in the fins or change colour. In each case, the cleaner fish understands the message, swims up to its client and over its body, into its mouth and out again through the gill covers, swiftly and nimbly (Ill. 144, 145). It removes parasitic fish-lice and isopods from the body and remnants of food from the teeth. Certain species of fish act as cleaners only as juveniles or occasionally. As a juvenile, the small wrasse, *Diproctacanthus xanthurus,* feeds by cleaning, but as an adult, nibbles at coral polyps. The Bluehead Wrasse, *Thalassoma bifasciatum,* cleans occasionally as a young fish, but is a minor predator, particularly of spawn, when it is adult. Representatives of Angelfishes, Butterfly-, Surgeon- and Damselfishes and others have been observed as juveniles engaged in cleaning activities. For fishes in coral reefs, the obligatory cleaners in particular are extremely important. When they were experimentally removed from a small reef, most of the other fishes moved away or suffered severely from parasitic infestation and fungoid attack.

Creatures of the high seas also have cleaners. Suckerfishes *(Remora and Echeneïs)* have a flat head disc that has evolved from the first dorsal fin, by means of which they attach themselves

145 111 112
113

111 *The very rare Four-spotted Butterflyfishes* (Chaetodon quadrimaculatus) *are found in pairs; their distribution extends across part of the Polynesian archipelago. (Society Islands, Pacific)*

112 *The Orange-striped Butterflyfish* (Chaetodon ornatissimus) *prefers to live at depths of up to 15 m in the calm waters of well-protected parts of the reef. (Society Islands, Pacific)*

113 *Equally brilliantly coloured are many marine flatworms, here* Pseudoceros affinis. *This species is quite common on coral gravel during the day. (Hawaii, Pacific)*

146/147 114 115 | 118
116 117 |

114 *A magnificent predator with pointed teeth inclined slightly inwards, the Coral Trout* (Cephalopholis miniatus). *(Red Sea)*

115 *This Moray eel* (Lycodontis undulatus), *thick as a man's arm, is disturbed by the diver's presence and extends its slightly opened jaws with the dangerously sharp teeth towards him in warning. (Red Sea)*

116 *The crab* Charybdis orientalis *goes in search of prey at night. Here it is eating a brittlestar. (Hawaii, Pacific)*

117 *With rows of powerful, chisel-sharp teeth, large Parrotfishes* (Scarus sp.) *noisily bite off and crush chunks of stony coral from the reef. (Red Sea)*

118 *The arch-enemy of stony corals is the Crown-of-thorns starfish* (Acanthaster planci). *(Society Islands, Pacific)*

148 119 |
120 121 |

119 *The small Fusiliers* (Caesio sp.) *need have no fear of the large Striped Madagascan Grouper* (Plectropomus maculatus) *except during its strictly observed feeding times. (Red Sea)*

120 *These small percoids remain close together in a school of about twenty individuals among the luxuriant growth of corals. Is their "warning coloration" an indication of their poisonous nature or merely a juvenile marking? Much still remains to be discovered. (Red Sea)*

121 *Small Grunts stay together under umbrellas of coral in shoals, here Bluestriped Grunts,* Haemulon sciurus. *(Southern Florida, Florida Strait)*

144

firmly by suction to sharks, rays, parrotfishes, turtles and even to divers and the hulls of ships, and allow themselves to be carried over considerable distances. They probably also search for and remove parasites from their "carriers". But they benefit mainly from food leavings. Their instinctive urge to attach themselves is so strong that the natives of tropical islands tie lines round the tail roots of Suckerfishes that have been caught, and return them to the sea. If they find a turtle, they attach themselves to its shield. The fisherman pulls in the line, together with the sucker and the turtle.

In addition to the important endosymbiotic association between hermatypic stony corals and zooxanthellae, true symbiosis exists between the small solitary Scleractinians *Heterocyathus aquicostatus, Heteropsammia michelini* and *H. cochlea* and the sipunculid worm *Aspidosiphon corallicola.* Their mutual association begins when a young sipunculid worm takes over the shell of a snail, perhaps that of a small Spire shell (Rissoidae). A coral larva also attaches itself to the shell. The coral grows, encircling the snail shell and the worm. The worm ensures that there is a clear channel inside the coral skeleton and on its underside, an opening that leads to the outside. It crawls forward, dragging the coral with it (Fig. 40). The worm is well protected, the coral will never silt up. Although the two coral genera belong to different systematic units—*Heterocyathus* to the family Caryophylliidae, *Heteropsammia* to Dendrophylliidae—they have evolved independently an analogous, beneficial and highly specialized way of life.

And what about the markings and colourings of fishes and other reef inhabitants? Butterfly-, Angel- and Surgeonfishes and other diurnal species are particularly brightly coloured and exhibit large areas of colour. These bold "poster-paint" colours stand out even in the frag-

mented reef (Ill. 93, 106, 111, 112, 122, 131, 136). This conspicuous coloration is not dangerous for the fishes, because large predators hunt only at twilight and at night after the colourful fishes have withdrawn to their dark sleeping quarters. Moreover, they always remain close to rocky ground that offers thousands of hiding places, into which they can disappear at any time. With their different food demands, many species are able to exist together peacefully. They defend their territory and the food resources it contains, only against competitors. Neighbours never eat the same food. So it is an advantage for a fish to indicate by bright coloration that it poses no competitive threat, since thereby it evokes no defensive reactions on the part of other fishes. Such activity is directed primarily at members of the same species (see p. 141).

Juvenile fishes would be unable to withstand these attacks that are often violent. Therefore their markings are different and they are not regarded as food competitors by adult fishes of the same species (Ill. 138, 139). And indeed, the food they eat is different. So the main effect of bright, bold colours is the preservation of the species.

In groups of fishes within a territory, those highest in rank are marked out by more intensive colouring. The dominant male in a shoal of Jewelfishes *(Anthias)* shows a trace of mauve or blue. Among Bluehead Wrasses *(Thalassoma bifasciatum)* the male has a blue head, the females and young have yellow and blackish markings. During the day, Fusilierfishes have strongly marked, longitudinal stripes that are visible through water for a considerable distance (Ill. 119). Particularly in the mornings when the fishes leave their resting places that are distributed widely through the reef, the markings enable them to assemble rapidly into a protective shoal. The territorial Damselfishes *Chromis pembae* that live in associations in deep water, have caudal fins that are almost white. In *Chromis dimidiatus,* the dark-coloured forepart of the body is in striking contrast to the light hind part (Ill. 132). As the small fishes whisk to and fro, they look like flashing signal lights. The ultimate effect is to hold the group together. Black-bordered caudal fins, such as those of the Damselfishes *Chromis nigrurus, Ch. ternatensis* and *Ch. simulans,* of the Sergeant Majors *Abudefduf sexfasciatus* and *A. coelestinus* and of the roving Fusiliers send out pursuit signals, causing fishes to assemble in a shoal or to swim one behind the other (Ill. 86). Genuine luminous spots are rare in reef fishes. The nocturnal Lanternfish *(Photoblepharon palpebratus)* has a crescent-shaped organ containing phosphorescent bacteria beneath its eyes that is highly luminescent. The fish can reduce the luminosity at any time by means of a pigmental shutter. By rhythmical activation of this device, the fish can emit flashing signals of about 0.2 seconds duration, that inform members of its own species of its presence. When it is pursued by an enemy, it "dips its lights" and disappears in the darkness.

Fishes equipped with dangerous specialized weapons give warning of their approach by optical signals. The venomous Firefishes are conspicuously coloured and so grotesquely shaped with their permanently extended, exceptionally long fins, that their appearance alone acts as a deterrent (Ill. 135, 137). The highly venomous Stonefish, *Inimicus filamentosus,* is so well camouflaged that it blends into the background, but the pectoral and caudal fins are coloured orange to salmon-pink. When it spreads them, they flash out a warning signal. In the same way, the Triggerfish *(Hemibalistes chrysopterus)* uses its orange-brown, white-edged caudal fins in a threat display when defending its territory. As a means of defense, Surgeonfishes have bright yellow caudal fins that are normally kept folded (Ill. 108). Perhaps they also warn

others not to approach. When Surgeonfishes defend themselves, they can inflict severe damage on the opponent by powerful strokes of the tail, with its extended "scalpel". Their habit of living in a group provides such a high degree of security that the defensive drive is suppressed. Grunts threaten both members of their own and of other species by opening the jaws wide. In this way they display the red colouring inside the mouth cavity. Rival grunts swim at one another with jaws agape. It looks dangerous, but culminates merely in the jaws being held together and "measured". The fish with the wider gape is the victor (Fig. 44). Ritualized combat, perhaps in the form of brief, violent duels by threat are very common in the reef. The dominant member of the group establishes his position rapidly and without bloodshed, and constant reduction in numbers within the species is avoided. In principle, fishes do not engage in extended bouts of fighting, which distract attention from the environment and increase the danger of falling victim to predators (Fig. 43).

Many Butterfly-, Bull-, Emperor-, Angelfishes and Moorish idols have black eyebands that camouflage the eyes almost completely (Ill. 88, 111, 112, 122, 139). Simultaneously or independently, they have evolved a black spot, frequently bordered in a light colour, on each flank near the tail root. These appear so like eyes that the attacker takes the back of the fish for the front, and is confused when it makes off in the opposite direction. Many fishes are able to alter the eye patch. In interspecific duels by threat, the fishes keep the patch in sight. If it becomes paler, it indicates the submission of the opponent and no fighting takes place (Ill. 42). Longnose Butterflyfishes, *Chelmon rostratus,* have the counterfeit eye in the dorsal fin. As they turn, they spread the fin and enlarge the eye. The subordinate fish folds its dorsal fin. Towards evening, the

43 *Aggressive butting behaviour in Bullfishes,* Heniochus acuminatus. *(After Eibl-Eibesfeldt, 1964)*

44 *Duel by threat between Grunts (Haemulon). The fish with the wider jaw is the victor. (From a photograph by Eibl-Eibesfeldt, 1964)*

Butterflyfish *Chaetodon melanotus* becomes a blackish colour with white flecks. It is possible that in the dark, the white flecks have the effect of large shining eyes and put predators to flight (Fig. 45).

Against the background of refracted light and shade that characterizes the complex structures of the upper parts of a reef, markings rich in contrasts have the optical effect of blurring contours, as with zebras in steppe country. With their pattern of light and dark stripes, Moorish Idols *(Zanclus)*, Bullfishes *(Heniochus)*, Sergeant Majors *(Abudefduf)*, Spadefishes *(Chaetodipterus faber)* and certain Sweetlips *(Gaterin)* blend effectively with their environment. Even the powerful Barracuda carries distinct bands of light and dark colour on its muscular body while it sleeps. Fishes in a shoal are difficult to distinguish from above because of their dark, blue or greenish backs and from below, against the light, because of their light underparts. Damselfishes, with their monochrome colouring of blue, green or grey are well matched to the colour of the water (Ill. 123). The intense red colouring of nocturnal Sweepers *(Pemphris)*, Soldier fishes (Holocentridae), Bigeyes *(Priacanthus)*, also known as Bullseyes and Catalufas, and of many crustaceans is the best possible camouflage (Ill. 82, 128). In dark caves or at depths of more than 10 m, the creatures appear black because of the absence of the red components of light in that environment (see p. 38). Fusiliers also adopt red coloration at night as a "hood of invisibility".

Predators as well as prey compete to achieve the best possible imitation of insignificant and therefore unnoticed objects in their surroundings—the first in order to approach as closely as possible to the prey, the second to escape notice. They do so by skillfully adopting a disguise (see p. 138). The effect is known as mimesis. The colour adaptability of Flatfishes (Pleuronectiformes), Octopuses and Sepias is a constant source of surprise. Light and dark colourings are achieved by expansion and contraction of melanophores brought about by endocrinal action. As a result, a Flounder appears almost black against a dark rocky ground, speckled against shingle, and light against a sandy ground, and can scarcely be distinguished from the substrate. The Cornetfish *(Fistularia)* has a clear, translucent colouring, and although up to a metre long, is only as thick as a thumb. As it slowly approaches prey frontally, it is impossible to distinguish. When it strikes, it is already too late for the unsuspecting victim.

Many organisms resemble plants and animals that live in the environment. Scorpionfishes (Scorpaenidae) that lie stoically immobile on the sea bed, and Stonefishes *(Synanceja)* and Frogfishes (Antennariidae) that conceal themselves in seaweed, have body appendages that look like algae (Ill. 142). Among Nudibranchs, *Melibe bucephala* lives on alga, *Phestilla melanobrachia* and a small yellow Staircase snail of the genus *Epitonium,* on orange-red Tube corals *(Tubastrea)*. All three are adapted to either algae or Tube corals in their coloration and skin processes (Ill. 141; Fig. 46). As the Porcelain snail, *Cyprea carneola,* crawls along, it covers its shell with its mantle on which there are many algae-like fringes. The Cornetfish (Ill. 110) aligns its slender, elongate body to the branches of the sea fans among which it lives and adopts their colouring, so that it is almost invisible. Many crabs and shrimps living on sessile organisms imitate their colour and structure and can scarcely be distinguished from them.

A series of vagile animals simulate not only the form and colour of other animals but also their behaviour in a remarkable way. The phenomenon is known as mimicry. It brings considerable benefits to those that practise it, and is the result of a very long process of adaptation. In addition to cleaner mimics (see p. 138), there are others living in reefs. Goatfishes or Surmullets, *Mulloidichthys mimicus* that make easy prey for predators because they have small scales and few spines, mimic the size, shape, colour marking and behaviour of the aggressive Snappers, *Lutjanus kasmira,* so exactly that in a mixed aggregation, the two are

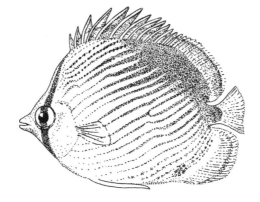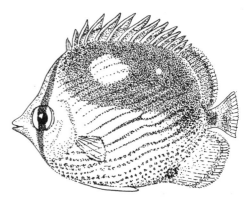

45   The Striped Butterflyfish, Chaetodon melanotus, *in its diurnal (left) and nocturnal (right) coloration. (Original)*

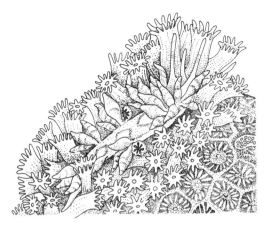

46 *Because of its body appendages, the nudibranch* Cuthona poritophages *is virtually invisible among the polyps of the coral* Porites lobata. *Natural size 8.5 mm. (After Rudman, 1979)*

virtually indistinguishable, and the mullets are not attacked. Young Red Snappers or Two-spot Sea perches *(Lutjanus bohar)* mimic harmless plankton-feeding Damselfishes *(Chromis ternatensis, Ch. iomelas* and *Ch. margaritifer)* and by swimming in a shoal of them, manage to get close to their prey. Juvenile Sea Basses of the species *Anyperodon leucogrammicus* are similar in appearance to harmless wrasses of the species *Halichoëres biocellatus,* and join up with them to approach prey. Small Cardinalfishes, *Siphamia argentea,* have developed the practice of collective mimicry. They always swim in a shoal close to bulky sea urchins of the genus *Astropyga* that have a similar dark colouring. When alarmed, they congregate in a bunch just above the sea urchin, in convincing imitation of its form. This affords them good protection. But they must beware lest they are pierced by the spines of the sea urchin. It would push them into its mouth using its pedicellariae, and consume them tail first. But without the sea urchin, the Cardinalfishes would have instantly fallen victim to the predator. The dark body of the innocuous and eminently edible *Calloplesiops altivelus* is spotted with bluish-white flecks, similar to the markings of the venomous Moray

eel, *Gymnothorax meleagris.* When attacked, it makes a threat display by imitating the front end of the Moray. The blenny *Meiacanthus lineatus* and related species, are rejected by predators because they bite and secrete unpleasant substances. Blennies of the genera *Petroscirtes* and *Plagiotremus,* certain species of Spine-check *(Scolopsis)* and the Cardinalfish *(Cheilodipterus zonatus)* protect themselves by mimicking the shape, colour pattern and behaviour of *Meiacanthus* blennies, and so predators assume that they are inedible.

Many animals are not eaten because they secrete unpleasant-tasting substances or because they have developed toxins. Workers observed that leatherstars are eaten by fishes only extremely rarely, and were able to isolate polyketid monoesters of sulphuric acid from the bodies, that proved to be an effective repellant. In tests, food to which no more than 0.2 per cent had been added, was refused by fishes. Certain nudibranchs produce acidic defensive secretions. They too are rejected by predators. Many of them have poisonous or unpleasant-tasting tissue. Other nudibranchs swallow unexploded stinging cells while they are grazing on coral polyps. These find their way into the blind sacs of the highly-branching midgut, that terminate in the dorsal appendages serving as gills. Since the stinging cells maintain their potency, they provide an effective means of protection for these otherwise defence-less sea slugs. As soon as a fish has taken what it assumes to be a tasty morsel into its mouth, it is stung and spits it out again. The number of poisonous sponges in tropical seas is increasing markedly. In one reef in the Gulf of Mexico, 27 out of 36 species of sponge are poisonous (Green, 1977). The markings of venomous Puffers (Tetraodontidae) are so conspicuous that enemies are warned off from afar (Ill. 127, 151). Many bony fishes have toxic glands. When threatened, the

"Moses Flounder" *(Pardachirus marmoratus)* secretes a milky poison made up of several components from a large number of glands situated along the dorsal and anal fins. It is fatal to small fishes even when diluted to one part in five thousand. Sharks that attempt to eat the flounder, experience paralysis of the jaw and take flight. A barracuda coming in to the attack at full speed is literally stopped in its tracks. Nevertheless, the Moses flounder is a popular item of food among the native population since the poison is destroyed when the fish is cooked. Firefishes, Stonefishes and Scorpionfishes have channels in the spiny rays on the fins and in the spines of the operculum that are linked to toxic glands (Ill. 135, 137, 142) and can cause a swimmer severe pain, cramp and paralysis and may even prove fatal. These creatures are sensitive to light and use their spines against any attacker casting a shadow. The long spines of the Hatpin sea urchin are very fine. A swimmer, thinking he is keeping a sufficient distance from them, may yet feel their painful burning sting. The tips of the spines are furnished with barbs and are very fragile, breaking off immediately in the wound. Since they are also hollow, they are difficult to remove. Within about a week, they are absorbed into the body. Most sea urchins have forceps-like pedicellariae that they turn on an enemy. With them, they inject a poison that induces respiratory distress and paralysis.

Other organisms make use of mechanical defensive devices. When threatened, *Gonodactylus guerni,* one of the smaller Mantis shrimps, closes the entrance to the cavity in which it lives with its curving, powerfully spined caudal plate. The Yellowhead Jawfish *(Opistognathus aurifrons)* barricades its dwelling tube in the evening by covering it with a stone. Triggerfishes (Balistidae) and Filefishes (Monacanthidae) retreat into narrow holes and cavities when alarmed. They raise the long powerful ray of the first dorsal fin, that is provided with a locking device, and force it into the rock so firmly that they cannot be withdrawn. Large Snappers also wedge themselves rigidly into rock cavities by means of their extended fins. Porcupine fishes inside their refuges swallow water and swell up like footballs, their spines stand out rigidly and it is impossible to take them as prey (Ill. 133).

Even the apparently defenceless Sea Cucumbers have developed a series of mechanical weapons of defense in addition to body venoms. Certain species extrude a sheaf of pinkish-white secretions which, on contact with water, become viscous elastic strands that entangle the molester. These Cuvierian tubes can be stretched to 20 or 30 times the original length before they snap, and it is no easy task for an enemy to disentangle itself from them. Still more astonishing is the sudden spilling out and ejection of the viscera from the anus, brought about by powerful muscular contractions, as the Sea Cucumber's ultimate defensive strategy. Its tough body covering lives on and regenerates the eviscerated organs within a few weeks or months. The metre-long Worm cucumbers *(Synapta)* are able to split themselves into several parts. But only the front end with tentacles is capable of

regeneration into a complete animal, while the remaining parts are sacrificed to the enemy.

Protective and defensive behaviour plays an important role in the preservation of the species. It is directed primarily against predators and food competitors. It is frequently linked with organs that serve in the process of food-getting, or equally frequently with body markings, defensive weapons and specialized patterns of behaviour. Interspecific rivalries are ritualized. The vanquished animal is scarcely ever killed because it indicates submission in good time, and the gesture of submission is respected by the victor.

Courtship, sexual activity and procreation

In the propagation of coral reefs, asexual reproduction of the sessile calcareous skeletal organisms by budding is predominant. It is particularly widespread among organisms that require only a narrow range of genetic variation for their survival. They include almost all sessile organisms. But in addition, corals, bryozoa and algae reproduce sexually, as we have already seen in the case of Scleractinians (see p. 17). This is important, because as a result of the fusion of egg

and sperm cell, genetic traits are strengthened and adaptive potentialities are increased. The reproductive drive is controlled by hormones, which in turn are activated by specific environmental factors. In the earth's temperate zone, young animals are born during spring and summer, when there is no lack of food. In coral reefs, there is no great difference between summer and winter, and certain organisms procreate throughout the year. But even here, an annual rhythm becomes more and more obvious as the distance from the equator increases, and many animals show a distinct reproductive periodicity. Land crabs return to the water only once in a year, and that for the purpose of propagation. At such a time, thousands of them can be seen making their way on their annual pilgrimage from an inland habitat to the coast.

Palolo worms, known to the Polynesians as "Mbalolo", are large marine annelids. Because their reproductive organs are considered a delicacy and because they appear in millions at predictable times, they have considerable importance as food organisms. The worms live in large numbers in channels and tunnellings in the coral reef, the Florida palolo *Eunice fucata* that is up to 70 cm long, off the tropical coast of North America, the Pacific palolo *Eunice*

47 As part of the courtship display, male Damselfishes perform remarkable leaping movements which both indicate the possession of a territory and stimulate sexual readiness in the females. (After Fricke, 1976)

viridis that grows to 40 cm, mainly in the reefs of Samoa and the Fiji Islands, and the Chinese palolo, *Ceratocephale ossawai,* off the coasts of Japan and China. Influenced by the phases of the moon, male and female sexual products mature in the long, thin-walled, chambered appendages that develop at the front end of the body in *Ceratocephale* and at the hind end in *Eunice.* At a particular time, segments containing eggs or sperm break away and swim up to the surface of the sea with writhing, spiralling motions. There they burst open, pouring out vast quantities of eggs and sperm into the ocean, turning the water cloudy and milky-white. In this way, fertilization and numerous progeny are ensured in spite of the many predators that gather for the feast: a biological benefit that is achieved only through the communal release of sexual products. The Chinese palolo spawns in October and November following the new and the full moon, the Florida worm before the last quarter of the moon, between June 29 and July 28, and the Pacific palolo in October and November, again after the last quarter of the moon. Palolos of the Malayan Archipelago swarm in March and April, and in the region of the Gilbert Islands, in June and July, in harmony with the phases of the moon. The hour at which the spectacle begins varies, getting later as one moves from west to east. If swarming begins at midnight off the westernmost Samoan island of Manua, it begins at 0.30 hours off Tutuila and not until 5 or 6 o'clock in the morning off the islands of Upolu and Savaii that lie farthest east. The phenomenon is dependent upon the tide, and lasts for only one or two hours. Since the intensity of spawning varies, the Fiji Islanders call the October swarm "Mbalolo lailai" or "small palolo" and the November swarm, "Mbalolo levu" or "great palolo". When it occurs, they go out in their boats to the reef or wade on the reef crest. They scoop the writhing morsels in delicately woven baskets or nets from the churning waters seething with countless bodies. Raw or rolled in leaves and baked, palolos are delicious to eat.

Most of the scientific research carried out so far on the reproductive biology of reef animals has concerned fishes. Aggressive rivalry alternates with courtship display, spawning procedures and brood-care patterns. The behaviour of solitary species is relatively simple, leading directly from the meeting of the sexes to the pre-spawning preliminaries. In the case of the Hamlet, *Hypoplectrus nigricans,* male and female swim up towards the surface as the sun is setting. Spawning occurs with the tail of one partner clasped round the neck of the other. Fishes with very different markings and which are considered as different species have been observed to engage in pre-mating activities, but no satisfactory explanation for this exists.

On the other hand, Jawfishes (Opisthognathidae) that usually live alone in tubes, show very complicated broodcare behaviour. They permit a partner to enter their living space only for the purpose of reproduction. Then the spawn is held in the spacious mouth cavity of the male and protected there until the young fishes have hatched. Mouthbreeders, that have long been familiar in tropical inland waterways, also occur in the ocean.

In those species in which a male maintains a harem, the male is constantly engaged in behaviour to indicate its dominant status and in guarding activity. The Cherubfish *Centropyge interruptus* establishes a small harem of up to four females. It keeps them together by swimming briskly round them. The circling and tail-beating movements provoke similar movements in the females, after which the bodies are brought into close contact and spawning occurs.

Among fishes living in larger groups, rivals fight between themselves to establish superiority. In many species, ritual threat displays are sufficient to determine the position of power. In a community of Tube eels (see p. 96) in which the tube dwellings of two males lie close together, the two will certainly compete as rivals. They incline themselves towards one another, raise the dorsal fins and stab at the adversary with open jaws until the weaker of the two quits the field. During the reproductive season, the male and the smaller female live in a single burrow. Their bodies entwine. The eggs are released into the open water and fertilized there. Huge Javanese Morays, *Gymnothorax javanicus,* were observed in May in a similar behavioural phase off the Society Island of Moorea in the Pacific. The cavity in which they lived lay at a depth of 34 m at the foot of a reef, and from it, a male and female kept watch, their erect bodies standing about a metre high, closely entwined. One of the two was somewhat smaller, but both were thick as a man's thigh and certainly more than 1.5 m in length. Communal living and entwining of the bodies were obviously also part of their reproductive ritual.

A complicated social structure has been observed in Jewelfishes *(Anthias squamipinnis)* (Ill. 109). Large males defend their territory. They spread the pectoral fins in such a way that a pale violet pattern is revealed—a means of emphasizing their masculinity and superiority. A second group is composed of non-territorial males that swim in shoals close to the ground. The third part of the *Anthias* population consists of females and young. They swim above the swarms of non-territorial males. Spawning time is introduced by the territorial males by means of dance-like movements. Eggs are deposited in open water.

An interesting phenomenon that occurs in the course of the life of many fishes is sex reversal. *Anthias* females become males after about a year. In certain labroids, two types of functionally potent

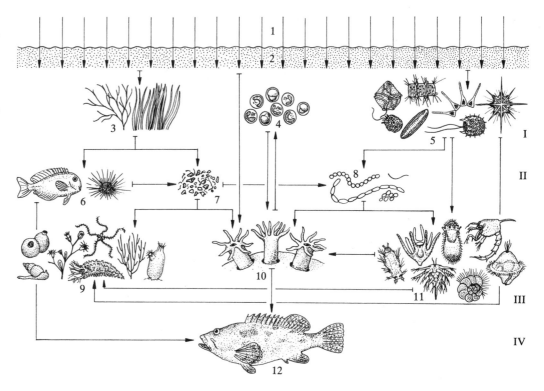

48 *Structure of relationships between the principal components of a coral reef biocoenosis. I. producers, II. plant consumers and destroyers, III. animals feeding on detritus and small organisms, IV. predators.*
1 solar energy, 2 dissolved mineral substances, 3 major algae and sea grasses, 4 zooxanthellae, 5 phytoplankton, 6 herbivores, 7 detritus, 8 bacteria, 9 benthos, 10 corals, 11 zooplankton, 12 predators. (Original)

males can be observed. Primary males are those that were originally masculine and remain so. Secondary males have developed from females following sexual modification. Since the process is associated with a particular body size, secondary males are larger than primary males and their colouring is quite different. This, in conjunction with the various juvenile colorations and female patterns that are different again, makes it no easy matter to assign different phenotypes of the same species. Moreover, there are behavioural differences. The primary males are group spawners, but the more colourful secondary males always mate with a single female. In the Cleanerfish, *Labroides dimidiatus,* sex reversal has been observed following the death of the male leader of a harem. In this case, the dominant female exhibited typical aggressive male behaviour, and within a short time, changed into a male. Within 14 to 18 days, the ovaries develop into functionally efficient testes capable of sperm production. The sexual change is very advantageous; large numbers of

females create many offspring but males come into being only as required.

Among Damselfishes (Pomacentridae), social relationships vary greatly. 31 species found in the Red Sea live either in solitary territories, in group territories with mixed sexes, in harem territories or in pairs. Some species alter the social structure in accordance with the supply of food and availability of hiding-places within the territory. Reproductive habits are also very variable. The male of the Blue Damselfish, *Chromis caeruleus,* spawns with 20 to 30 females. The spawn is deposited in nests of green algae. The nests are guarded by the male for two days until the larvae hatch. The Sergeant Major *(Abudefduf zonatus)* spawns in empty musselshells or in holes in the rock. The male keeps watch over the eggs and fans a supply of fresh oxygen-rich water towards them. It uses its mouth to remove sand that has been washed in by wave action caused by the violent fanning of its fins. In *Acanthochromis polycanthus,* brood care extends to allowing the young fishes to feed on

the skin of their parents and sometimes also of other fishes, for a considerable time after they have hatched.

Grey Humbug *Dascyllus marginatus* maintain a harem among densely-branching Styloid or Spadiciform corals. Large colonies with extensive branching can house several harems. If space is restricted, the inferior males are driven away. They represent, in any case, a burden on the natural resources and soon fall prey to predators. The reproductive ritual is initiated by the dominant male in a long sequence of signalling leaps, a form of swimming incorporating jumping movements in which growling sounds are emitted at the highest point of the leap immediately before the fish pitches downwards again (Fig. 47). In this way, it marks off its territory against intruding males and stimulates the females of its own as well as of neighbouring harems. Sometimes in densely colonized habitats, the males display to females of lower rank in neighbouring harems. In spite of vigorous retaliation on the part of the dominant males there, it is sometimes possible to induce one of them to defect. Females of lower rank are clearly not so closely bound to the corals in which they live as are those of high rank, and to prevent them from moving into a different harem, they must be constantly courted by display. On spawning days, the largest male is dominant even over the smaller harem-leaders in the immediate vicinity; all the females swim to him to spawn. The eggs are attached by means of mucous filaments to rocky ground cleared by the males, and are then guarded and fanned by the males. The largest female spawns first, to be followed by the others in order of rank. In the evening, two or three days later, the young fishes hatch. This is the most favourable time of day: in the dark, plankton-feeders are unable to locate the tiny creatures directly, and losses are relatively slight.

The Anemone Clownfish (see p. 143) also shows complicated reproductive behaviour (Ill. 125). *Amphiprion bicinctus* live in pairs in an anemone. The couple remains together for years. Before depositing the eggs, both partners pluck at the ground, push the bellies together and vibrate their bodies until the reproductive products are given off. More than a dozen times a year, the female attaches her eggs to the rocky ground close to the anemone in which they live. The male fans fresh water over them for ten days, until the young hatch out. In *Amphiprion akallopisos,* the larger female is the head of the family. She maintains a harem of up to 11 males. The strength of the individuals determines a strict order of rank. If the female dies, the highest ranking male becomes a female. In fishes, sex reversal is subject to social control. The more specialized the species, the more complicated its reproductive biology. The specialization of Amphiprions in respect of food and habitat has obviously been accompanied by specialization in reproduction.

Collective harmony

What monotony and paucity of species marks the civilizations built by man, in spite of their great productivity, when compared with the tropical jungle on land and the coral reef in the ocean! A consideration of the previous chapters must make this obvious, even though they can never come near to reflecting the great diversity of species and forms of adaptation that exist in a reef. Yet perhaps they may sketch in outline the way in which the multiform complex of the coral reef has developed over millions of years, in which hermatypic Scleractinia and other organisms, in symbiotic association with algae, have transformed the calcium dissolved in the ocean into rock, and in which thousands of organisms have adopted the resulting substratum for their own use as habitat and source of food (Ill. 3, 4, 123). Wave action, surf and currents ensure the necessary exchange of oxygen and carbon dioxide, of nitrogen and phosphoric compounds necessary for metabolic processes. The growth and propagation of algae provides the basic element in a feeding network that extends from symbionts to predacious carnivores (Fig. 48).

Tons of material in solution are bound in the ocean every day by living organisms. But they are all subject to a natural cycle as a result of which the pattern of distribution does not alter. For all living beings die, decay, undergo remineralization and are once again reduced to the original chemical components. The total of calcium carbonate that is bound and dissolved in great quantities is supplemented from the continents by way of the rivers. But in the final analysis, even this material has its origin in marine sediments that became part of the terrestrial region of the earth as a result of land elevation in the grey mists of pre-history.

The abundant supply of highly diverse habitats and food resources has led to a

161 126
127 128

126 *The Lagoon Ray* (Taeniura lymna) *with poisonous spines in its tail. (Red Sea)*

127 *The Black-spotted Blowfish* (Arothron diadematus) *pumps itself up into an irregular sphere when provoked. (Red Sea)*

128 *The Bigeye* (Priacanthus hamrur), *an animal of the night. (Red Sea)*

162/163 129 130 | 135 136
131 132 | 137 138
133 134 | 139

129 *The Red Wreckfish* (Cephalopholis coatesi) *is a predator of the reef. (Red Sea)*

130 *The Spanish Hogfish* (Bodianus rufus) *is a minor predator. (Southern Florida, Florida Strait)*

131 *The Maskarill-Butterflyfish* (Chaetodon semilarvatus), *an endemic fish of the Red Sea.*

132 *Between the small Pore coral* Synaraea convexa *swims the Butterflyfish* Chaetodon reticulatus. *(Society Islands, Pacific)*

133 *This Spiny Boxfish* (Chilomycterus echinatus) *is able to wedge itself into holes. (Red Sea)*

134 *The extended snout of the Longnose Butterflyfish* (Chelmon rostratus) *allows it to extract small worms from narrow crevices. (Red Sea)*

135 *The Spiny Firefish* (Pterois radiata) *in a cave. (Tuamotu Archipelago, Pacific)*

136 *Remarkably colourful Queen Angelfish* (Holacanthus ciliaris). *(Cuba, Florida Strait)*

137 *This rare Zebra Firefish* (Pterois russelli) *is characterized by the high toxicity of the glands. (Tuamotu Archipelago, Pacific)*

138 *The young Imperial Angelfish* (Pomacanthus imperator) *with blue and white semicircles. (Red Sea)*

139 *Sexually mature Imperial Angelfish* (Pomacanthus imperator) *with yellow stripes. (Red Sea)*

164 140 141 |
142 143 |

140 *Hermit crab enclosed by a sponge. (Red Sea)*

141 *The nudibranch* Phestilla melanobrachia *(right) is feeding on Tube coral* (Tubastrea coccinea). *It is well adapted in shape and colour to the organisms on which it feeds. (Hawaii, Pacific)*

142 *Scorpionfish (*Scorpaena sp.). *(Red Sea)*

143 *Wrasse of the genus* Cheilinus *with signal spots on the back. (Red Sea)*

144 *A Cleaner fish (Labroides dimidiatus) removes parasites and organic dirt particles from a large Parrotfish (Scarus sp.).*

145 *The Cleaner fish (Labroides dimidiatus) has started its cleaning activity on a small Neon Abudefduf (Abudefduf oxydon).*

146 *A symbiotic relationship exists between this large nudibranch (Hexabranchus marginatus) and a shrimp which, in its colouring, is well-adapted to the living substratum on which it resides.*

147 *The Cleaner shrimp (Hippolysmata grabhami) removing dirt from a Moray eel (Lycodontis sp.).*

148 *Because lagoons have an abundance of food and plankton, they are a very favourable environment for bivalves. In some places, Tree oysters (Isognomon radiatus), also known as Coon or Mangrove oysters, that normally grow on mangrove roots, are cultivated commercially. (Virgin Islands, Western Atlantic)*

149 *Although the reefs of Southern Florida are protected, traders import shells and stony corals by the ton from other coral reef areas.*

150 *It is not known whether these small cairns built up out of layers of coral gravel on a lonely atoll are used by the native inhabitants as a navigational aid or in the performance of religious ceremonial. (Tuamotu Archipelago, Pacific)*

151 *Slow-swimming reef creatures in particular, such as this Black-spotted Blowfish (Arothron diadematus) will play readily with divers. (Red Sea)*

152 *In the dim light at depths below 20 m, where the Copper Snapper (Lutjanus bohar) is usually found, the white spots effectively signal its presence. (Tuamotu Archipelago, Pacific)*

dense concentration of living organisms in the coral reef. In the present state of scientific knowledge, it is not possible even to guess at their numbers. Although stony corals are the major contributers to the construction of the underwater limestone mountain ranges, it is rather the crustaceans, molluscs, ringed worms, round worms and fishes that make up the majority of the multicellular animal species. And what we know about their biology is still only fragmentary. Our knowledge of other phyla and classes of the plant and animal kingdoms occuring in coral reefs is even more meagre.

From an ecological point of view, opportunists and specialists are the organisms that determine events within a reef. The opportunists, or euryoecious organisms, can survive and flourish in the most varied environmental conditions, they exist virtually everywhere and have adapted to almost every situation, provided only that no extraneous factors have been introduced by man. They produce vast numbers of progeny in the form of gametes or larvae, most of which are consumed during their planktonic existence, but some of which survive and multiply. They include algae and stony corals, and only as such are they capable of restoring and renewing the biotope of the reef. The specialists, or stenoecious organisms, never form large populations of many individuals. They produce only enough offspring to ensure the continuation of the species, but are, on the other hand, so highly specialized that the danger of extinction is avoided. Because of specialization, they are fundamentally superior to the opportunists that survive by reason of their vast numbers, and moreover, they are capable of asserting themselves in face of the latter. The pressure of competition and selection that exists in the coral reef has led to the development of an extremely high proportion of specialists among the total number of species existing there.

The coral reef is a benthic community. Its organisms are either vagile or sessile, the sessile organisms either flexible or rigid. The larvae of most benthic organisms are planktonic. They are eaten as food or drift in sea currents. Only a fraction of them find a place to settle, live and grow into a sexually mature creature or colony. The opportunists find such a place quickly because of their high degree of adaptability. It is much more difficult for the more highly specialized species to find an appropriate home. Frequently they must move about for a long time, but they are able to live as planktonic larvae for long periods, often weeks or months. So the survival of all is secured.

The most outstanding feature of coral reefs is that they are constructed primarily by means of lime-producing animal-plant symbiosis involving corals and zooxanthellae, and that they are constantly restored and renewed in this way. Their autogenous reproduction is the basis of the large number of ecological niches and species. This reproductive capacity and the high total of species and individuals are to be interpreted as an inevitable evolutionary consequence of the pressure of natural selection, that came about on the basis of adaptive strategies and interspecific relationships that are not yet fully understood today. As a result, coral reefs are not only exceptionally beautiful but they also contain many ecological, evolutionary and biological secrets that call out for investigation. Most of it is still to be done.

Benefits and dangers

of coral reefs to man

It was only the development of free diving techniques that gave access to the diversity and colourful profusion of life in the coral reef. But neither love of nature nor modern technology would be sufficient motivation for the intensive investigations currently being carried out, without the added incentive of potential profit. To understand this, it is necessary to stress the importance of coral reefs, the damage caused by the intervention of man and the responsibility that devolves upon us.

Of all marine biotopes, the coral reef produces the greatest biomass. Many organisms useful to man as a source of food live in it: fishes, turtles, crustaceans, snails, bivalves (Ill. 48), sea cucumbers, palolo worms and algae. While their culinary use is generally not yet widespread in the more northern countries, they are particularly important in the diet of inhabitants of tropical reef regions, where there is an acute shortage of protein. Commercial cultivation of some of the organisms has already begun. Currently, experiments being carried out in farming algae, shrimps, rabbit fishes, mullets and turtles still suffer from our fragmentary knowledge of their biology and ecology, and there are many failures, but they are undoubtedly an indication of the way in which coral reefs could, in the future, be put to agricultural use. Under the pressure of the increasing world population, their use as a food resource will increase in importance in the coming years, and enhance the quality and quantity of the world's food supply. In addition, the continuous production of planktonic larvae in a coral reef makes a considerable contribution to the plankton stocks in adjacent parts of the ocean, so that in the long run, the huge shoals of economically important fishes such as sardines, mackerel and tuna fish benefit from it.

Raw materials for the pharmaceutical industry are obtained from certain coral reef organisms. For example, prostaglandin, a derivative of fatty acid, that produces a medically useful reaction in muscle, became known some years ago. Its clinical application covers a wide range, extending from the treatment of cardiovascular disease through asthma and gastric ulcers to its use in assisting childbirth or terminating pregnancies, without producing damaging side effects. The hormone was found in minute quantities in the seeds of asperifoliate plants (Boraginaceae), in the urinary bladder of sheep and other animals. At that time, prostaglandin was 100 times more valuable than gold. But then it was discovered in a Gorgonian coral common in reefs, from which it could be extracted not by the milligram but by the gram. Research scientists from the University of Oklahoma, U.S.A., carry out regular programmes of diving in search of marine organisms containing useful drugs. Now research is concentrated on sponges, sea anemones and corals.

During the millions of years in which coral reefs have existed, an enormous biomass has been converted by physico-chemical action into mineral oils. The oil is stored in the porous limestone of the reef and at the same time, continuously augmented, directly at its place of origin. So recent as well as fossil coral reef regions are potentially rich in oil. Numerous sites of oil discovery in Venezuela, Mexico, Texas, the Near East, the Arabian Gulf, the Arabian Peninsula, off the coast of Australia and in the shallow waters between South-east Asia and Australia are of marine origin (Fig. 49). The continual increase in biological and geological research subsidized by the petroleum industry indicates the enormous importance of reef oils as a source of energy. Competition in this field can even bring about military conflict, as wars in South-east Asia, the Near East and Africa have proved.

The average production of calcium carbonate in a coral reef varies from 400 to over 2,000 tons per hectare annually (Chave, Smith and Roy, 1972). Since coral limestone has a high degree of purity, it is much sought after as a basic material for the building industry, where it is used as building stone, mortar and cement. Whole villages, towns, breakwaters and port installations have been built of coral limestone. In some areas, coral debris (Ill. 150) is gathered up together with mollusc shells, crushed and ground for use as fertilizer.

One important aspect of living coral reefs is usually overlooked. Every day, the world's many rivers carry immense

170

quantities of dissolved calcium carbonate into the oceans. Over many millennia, concentrations would have increased leading to alterations in the chemical reaction mechanisms of sea water. But this was prevented by plants and animals that converted the dissolved limestone into skeletal elements and deposited them on the sea bottom, and not least among them were the stony corals. Thus the chemical properties of the ocean remained basically unchanged, and the world of the marine organisms was able to continue to develop relatively undisturbed in its favourable, stable environment until our own day.

Coral reefs are situated off continents and islands to which they provide an effective measure of protection. Since their topmost parts lie just below the surface of the water, the waves break on them, and the land behind escapes the onslaught of the sea (Ill. 65, 67, 81). The lagoon lying between the reef and the land is a calm-water zone with a high production of biomass and considerable silting, where filling-in by sedimentation is in progress. So coral reefs represent a potential sedimentation element, and there is no other biogenic factor that has influenced coastal formation as radically as they.

The uncontrolled upward growth of coral to the surface of the water represents a considerable danger to shipping, as evidenced by large numbers of wrecked ships on reefs. In extreme cases, even entire harbours on major trading routes can be rendered unusable by the proliferation of stony corals. In the Sudanese harbour of Suakin at about the turn of the century, the growth of corals inexplicably reached such massive proportions that ocean-going vessels were unable to put in there. Only after that was Port Sudan built. Suakin declined, and can be visited today as an open-air museum. So far, no means is known of preventing huge financial losses incurred in this way.

While traditional pearl fishing has been replaced to some extent by cultured pearl production, it is still practised in the Persian Gulf and in northern Australia. Pearl mussels are often found in coral reefs. Pearl cultivation is carried out successfully in atolls in the South Seas. Pearl oysters *(Pinctada)* are threaded on nylon lines, injected and left in the nutrient-rich waters of the lagoons until pearls have formed inside them.

As a delightful and fascinating environment for the tourist, coral reefs represent an eminently marketable product (Ill. 4, 151). Since the reefs are set in clear, warm waters and since today, aeroplanes can reach even the most remote islands rapidly, they have become economically profitable from this point of view. Today, the tourists who disport themselves in the ocean with diving mask, snorkel and flippers are numbered in millions. Glass-bottomed boats provide a clear view down into the sea, so that even older tourists are able to experience the world of the coral reef. Sensibly enough, whole tracts of reef in the U.S.A., Kenya, the Sudan, Australia and other countries have been declared areas of nature conservation, so that at least in some places, a limit has been set to the overhunting of fishes and the collecting zeal of souvenir-hungry tourists that was threatening coral and fish stocks. Unfortunately, lack of supervisory personnel makes it difficult to maintain effective control over protected areas. And since tourists like to buy souvenirs that have come from the sea—beautiful corals, shells and starfishes—business flourishes in spite of legal restrictions (Ill. 149).

The importance of recent stony corals and reef associations in basic research is considerable. Since as reef-builders they are dependent upon warm seas, their distribution enables the biogeographer to delineate the tropical region of the ocean. From recent corals, palaeontologists can draw valuable conclusions on the environmental conditions that existed when

they were living, and on the development and age of the formations. From an examination of day and night zones of growth, it has been possible not only to establish the age of fossil corals, but also to confirm the astronomers' claim that the earth rotated considerably more rapidly 400 million years ago than it does today.

A notable feature of stony corals is their great sensitivity to water pollution. If their degree of sensitivity is quantified, it is possible to predict the maximum amount of pollution a particular region of the sea can tolerate. Conversely, since the degree of sensitivity varies from species to species, it is possible to draw conclusions on the extent of water pollution from an observation of coral deaths. A check can be maintained without difficulty on corals since they are sessile organisms, and in this way they serve as a gauge of environmental pollution, an indicator of the state of health of the ocean (Fig. 50).

In the earth's tropical belt, in an area of sea measuring 123,000,000 km^2 and representing about a third of the world's oceans, the unceasing work of polyps has created thousands upon thousands of coral reefs. When they reached the surface of the water, banks of pebble and sand, silted-up lagoons and elevated sea beds developed into islands and atolls with evergreen forests and groves of palms. The coconut tree provides man with wood for building, fibres for clothing, milk and fruit pulp for food. Among the living reefs, fishes, crustaceans, mussels and turtles are caught. In this way, coral reefs provide a home and livelihood to many people (Ill. 150).

Some coral islands lie in the middle of the ocean, far from the mainland, in a situation of extreme strategical significance for certain military powers. For this reason they have been developed as military bases. Guam, Okinawa, the coral island of Kwajalein belonging to the Marshall Islands and many others bear witness to this. The Chagos Archipelago in the middle of the Indian Ocean is still a British colony. The atoll belonging to it, Diego Garcia, has been leased to the U.S.A. for an initial period of 50 years as a military support base. The inhabitants were resettled on Mauritius. The coconut-palm forest was bulldozed flat to make runways for heavy bombers. Dock installations for atomic submarines, arms and ammunition depots, fuel stores and harbourage for fleets of rocket-carrying vessels were created. American military forces recently established support bases on the island of Tinian to the north of Guam and the island of Masirah off the south coast of Oman, and these are certainly not the last of such projects. The French government used the Pacific atoll of Mururoa as an atomic test site. After several explosions, pollution of the atmosphere by radioactive material was such that in 1973, Australia and New Zealand temporarily broke off relations with Paris. The U.S.A. similarly misused the Eniwetok Atoll and the Bikini Atoll in the Marshall Islands for atomic tests 66 times between 1946 and 1958. Beforehand, the inhabitants were compulsorily evacuated "for a few months". The ravaged Bikini Atoll remained polluted with radioactive strontium 90 until 1980. Palmtrees planted experimentally turned yellow, sea birds and turtles produced no young.

Actions carried out thoughtlessly and in ignorance by men motivated by greed for profit continue to affect coral reefs, and have already caused irreparable damage. In many cases, inappropriate, ill-considered alterations to coastal regions have proved fatal to corals. As a result of destruction of plant cover, deforestation measures, burning-off of grassland, levelling and construction work in coastal regions, fertile topsoil is exposed and washed into the sea. Here it settles on reefs and chokes corals and other sessile organisms, as has happened

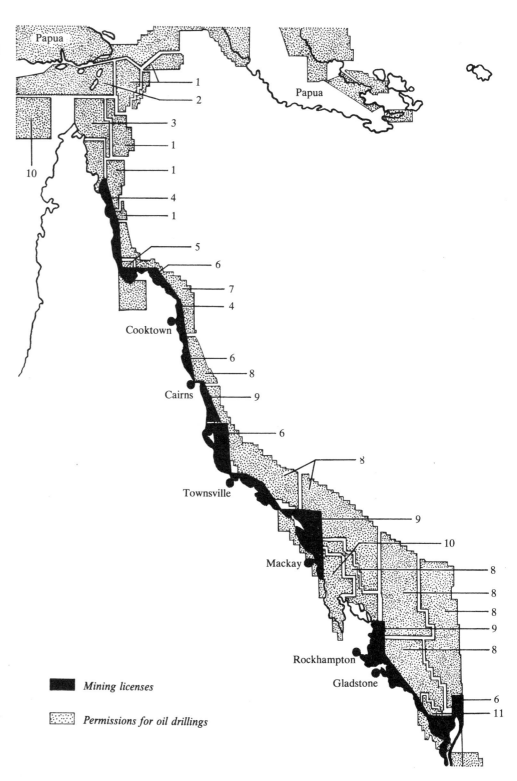

■ *Mining licenses*

░ *Permissions for oil drillings*

49   Concessions to oil producing firms for the area of the Great Barrier Reef were granted years ago by Australia. 1 Tenneco Aust. Inc. & Signal (Aust.) Pet. Co., 2 Calif. Asiatic Oil & Texaco Overseas Pet., 3 Gulf Interstate Overseas, 4 Eastern Prospectors P/L & Offshore Drillers P/L, 5 Exoil NL & Transoil NL, 6 Ocean Mining AG, 7 Corbett Reef Ltd., 8 Aust. Oil & Gas, 9 Planet Mining Co. P/L, 10 Ampol Expl. (Qld), 11 Shell Dev. Aust. (After Bennett, 1971)
Meanwhile the Australian government has withdrawn the licenses for mining and oil drilling.

off Molokai, the coast of Tanzania, the Seychelles and Hawaii. Human activity of this kind has caused both the loss of fertile soil with a reduction in agricultural potential, and destruction of reefs.

The action of dragging heavy objects across the sea bed destroys sessile and hemisessile organisms. Very often, trawlers fishing near coral reefs stir up fine sediments. Bottom sand and mud remain in suspension in sea water for up to five years before returning to the sea bed, and drift in quantity onto the reefs, where they settle. At the same time, suspensions reduce the oxygen content of the water. This again is detrimental to corals which can scarcely tolerate an oxygen content that falls below the normal nighttime minimum. Phenomena of this kind have been reported from Guam, Johnston Island, the Virgin Islands, Singapore, Samoa, the Seychelles and Hawaii. The use of dynamite in fishing, although to some extent prohibited, is nevertheless widespread in reef areas. Destructive exploitation such as this destroys entire fish stocks and future generations and also the coral reefs.

Sediments stirred up by the explosions settle on neighbouring reefs and damage the coral stocks. Similar destruction caused by dredging work and levelling of the sea bed has been reported from Bermuda and personally observed off Hodeidah. In the Gulf of Mannar, blocks of limestone and large living coral colonies are quarried and used by the building industry for the manufacture of cement. As a result of this exploitation, the reefs have been weakened to such an extent that in 1970, a cyclone drove the uncontrolled water masses raging across the reefs and onto the shore. Lives were lost, villages and roads destroyed. But this did not bring an end to the removal of reef material off Rameshwaram. Catastrophes have been brought about in the same way in Malaysia.

Everywhere, industrial waste products cause devastation. Hundreds of tons of leaves, stalk parings and pulp from sugar cane are washed into the sea every day from sugar refineries on the Caribbean islands and Hawaii. The water has become very cloudy, the number of coral and fish species is drastically reduced. In Hawaii's Kaneohe Bay, which was still unaffected in 1928, more than 99 per cent of the corals in an area of reef of 880 hectares were destroyed a few years ago by the introduction of municipal sewage and by erosion of the shore region. The reasons were the turbidity of the water, silting, increased concentrations of phosphates, the development of anaerobic strata and the formation of hydrogen sulphide. The coral skeletons were covered with algae and dirt and had a grey, desolate appearance. But now, since the site of sewage discharge has been moved a good distance to seaward of the reef, they appear to be recovering.

Eutrophication brought about by artificial pollutants also stimulates the growth of coralline algae, a phenomenon observed off Hodeidah. The presence of sewage causes abundant growth in polychaetes, various sponges and holothurians. In addition, organic enrichment has resulted in a reduction in the oxygen content that inevitably damages reef organisms.

The discharge of municipal and industrial sewage is frequently followed by reduction in salinity. Corals react with great sensitivity to changes of this kind, and a mixture of 25 per cent of fresh water and 75 per cent of salt water proves fatal to them within a week. A decrease in salinity also reduces lime production. On the question of municipal sewage, it is worth noting that the World Health Organization's directive on permissible pollutants in water at bathing resorts specifies a maximum of 350 faecal bacilli or a total of 1,000 coliform bacilli per 100 ml. But in the waters of a normal municipal sewage inflow point, 11,000 coli bacilli per 100 ml have been found, and off a discharge point from a hospital in Palau, as many as 34 million. In water from the harbour of Kingston, Jamaica, 240,000 coli bacteria per 100 ml have been counted (Wood and Johannes, 1975).

Highly aromatic oils produced by oil pollution usually have an immediately fatal effect. These highly volatile components of crude oil are released only on contact with water. They are absorbed relatively quickly by organic suspensions and so make their way into the food chain of marine animals. Heavy oil residues sink to the bottom. Investigations off Bermuda, the island of Japtan and in the Eniwetok Atoll have shown that they are broken down only slowly in marine sediments.

On the whole, the corals were not damaged by oil moving over them. But volatile oils and anti-oil detergents damage their tissue. In one case, a substance used to dissolve spilt oil was found to be one hundred times more poisonous than the crude oil itself. The oxidation of 1 litre of mineral oil by bacterial degradation requires 320,000 litres of oxygen-saturated sea water. So oil slicks must inevitably cause ecological damage to the environment. Layers of drifting oil harm reef invertebrates and seabirds as well as corals. This is particularly serious because extensive oil-drilling projects with their attendant high level of pollution have been planned or are already in progress off many sea coasts—in Southeast Asia, the Persian Gulf, in the Caribbean, off Fiji and Tonga. On the whole, these drilling operations were not preceded by essential ecological investigations. In the relatively small area of the Central American oceans there are more than 60 oil refineries. The seas are so badly polluted by drifting oil slicks and oil residues coming from hundreds of ships that have run aground, or which, in the interests of profit, have illegally discharged their tanks, that it can only be described as permanent and irreparable damage to the ocean and its organisms.

The danger to marine ecosystems as a result of undue heating caused by power stations and other industrial installations is particularly high in the tropics. In the Torres Strait, even the natural heating of shallow lagoon waters by the sun has prevented the growth of corals along an off-shore strip 150 m wide. Similar phenomena are found in the Ano Atoll, off Turky Point in Florida, off Guam and Hawaii.

Many countries and islands fringed by coral reefs suffer from a shortage of water. Arabian and African countries, the U.S.A. and others are therefore endeavouring to compensate for the deficit by extracting fresh water from the sea. The desalination plant necessary for this process not only increases the salinity and temperature in the area round about by the discharge of brine and cooling-water, but at the same time introduces poisonous metals such as copper, zinc and nickel. Their higher specific

gravity causes the effluent to flow directly across the sea bed. Although certain corals tolerate a salinity of up to 48 per cent, most of them are destroyed by increases in salinity. Gorgonians are even more sensitive. The interaction of various factors makes analysis of the damaging processes difficult. Only in one of eight cases were prognostic ecological enquiries carried out before the construction of a desalination plant.

The use of pesticides in agriculture to combat plant and animal pests and to increase yields is common throughout the world. The practice always has disastrous consequences, because not only pests but the entire organic world is affected. Herbivores store within their tissues chlorinated hydrocarbons, derivatives that are difficult to break down, and these are then passed on to microorganisms, predators and carrion-devouring animals. Because rain and river waters wash the pesticides into the sea, their distribution is worldwide; they have been found even in the liver of

Antarctic fishes. And of course, they can also poison humans. Many countries that have banned the use of pesticides because of the lasting damage they cause, nevertheless continue to produce them for export to other countries, so that even today, their use is still spreading. Little is known of the effect of chlorinated hydrocarbons on corals, but we do know that they increase respiration rates and decrease the rate of photosynthesis. Dangerous amounts of pesticides in lagoons and atolls usually come from neighbouring coastal areas. On 17 April 1979, they destroyed some 20 tons of fishes in the Truk Lagoon. Six local people who had eaten some of the affected fishes became seriously ill.

Atomic test explosions also have a directly destructive effect on coral reefs and marine organisms as a result of radiation. Inevitably the entire food chain within a radius of several hundred kilometres is affected by radioactive contamination, and in the case of the Bikini Atoll, it was 240 km. One serious conse-

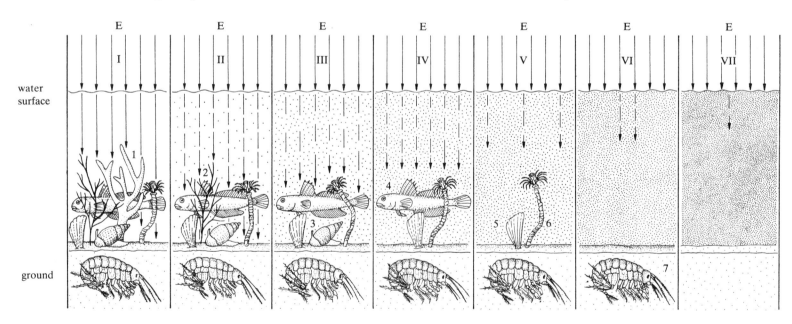

50 The greater the quantities of sediment, man-made waste, pesticides and other toxic materials that enter the ocean, the higher the death rate among living organisms (I–VII). In VII, only anaerobic bacteria that produce evil-smelling hydrogen sulfide are able to exist. E solar energy supply.

1 corals with zooxanthellae as zooplankton-feeders and primary producers, 2 algae as primary producers, 3 snail as herbivore, 4 goby as carnivore, 5 and 6 bivalve and tubeworm as plankton-feeders, 7 sea flea as detritus-feeder. The corals die first. (Original)

quence for local people is the closing of thousands of square kilometres of fishing grounds, for four years in the case of the Rongelap Atoll. Moreover, because of the complexity of atmospheric and oceanic currents, it is virtually impossible to predict the geographical spread of radioactive contamination. After a reef is destroyed, recolonization by corals occurs with a distinct reduction in the number of species, as was observed in the Eniwetok Atoll. In atomic tests in the Bikini Atoll, three islands were destroyed. The continuous testing of nuclear weapons has already devastated the Mururoa Atoll to such an extent that the possibility of its complete collapse and subsidence cannot be ruled out.

Blue-green algae are often the first organisms to colonize dead reef substrates and coral skeletons. Normally edible fishes that graze on the masses of algae, frequently exhibit a form of toxicity known as ciguatera. Anyone eating such fishes becomes ill and may die. Blue-green algae, Cyanophyceae, are believed to be the primary vehicles for the toxin. Bagnis showed that acceleration in the spread of ciguatera, as the coral reef is destroyed, is probably caused by increasing numbers of blue-green algae settling there.

The drastic reduction in reef organisms as a result of the commercial removal of corals, bivalves and snail shells, and by the taking of aquarium fishes is widespread off East Africa, in the Red Sea, in Malaysia, Indonesia, Tahiti, Florida, the Bahamas, the Philippines and elsewhere. Instead of legislation prohibiting their removal, which is ineffective because impossible to enforce, it might be better to impose effective restrictions on trade in coral, similar to those introduced for endangered species by the member countries of the International Trade Convention. The taking of corals for commercial purposes is forbidden in the Florida reefs. Nevertheless coral, probably from the Bahamas, is offered for sale by the lorry-load. Whole shipments of corals and shells are imported from the Philippines and from other Pacific reef areas for sale in the U.S.A. Shops that include marine organisms among their wares are found all over the world (Ill. 149).

Man who came into being out of unadulterated nature by a process of evolution, only three or four million years ago, is still today part of a natural totality, even though civilization and technology have lent him a certain independence and autonomy, and may lead him to shortsighted actions. Man will never be able to live without oxygen, water and food—that is self-evident. But oxygen, water and food have not always existed on the earth. They too came into being in the course of evolutionary development and are constantly renewed by nature: the oxygen in the air by the green plants that are on land and in water, water by the hydrological cycle of the oceans and by clouds that produce rain, food by the propagation and multiplication of living organisms.

If air and water are poisoned, all living creatures will die, including man. He knows this, and yet continues to pollute nature with poisonous exhaust gases, industrial waste, municipal sewage, pesticides and oil residues, with building rubble, dirt, ash and radioactive materials. Air and water carry all these deadly, destructive products to the furthermost corners of the earth. I have seen animals of many kinds in large numbers lying poisoned and dead on the sea bed or drifting in the water. I have seen how terns and gulls swooped down and fed upon them. And I have seen thousands of seabirds, themselves poisoned, dead on the surface of the ocean or cast up on the shore. Yet man continues to kill. The munitions industry in particular, in addition to discharging vast amounts of harmful waste, manufactures huge quantities of deadly products.

Man is not able to control nature. At best, he can learn to modify nature for his own benefit, in such a way that its basic structure and its ecological balance are not damaged. It is not yet too late for that, but man must understand and act rapidly to avert his own destruction and that of the entire delicately balanced biological structure of the earth.

Consider the coral reef! It is full of life, full of luxuriant growth, full of energy. It is richly diverse, entrancingly colourful and beautiful. In addition, it is useful to us. But it is endangered. While nature itself renders its existence precarious enough by raging cyclones, violent seas, harsh erosion and endless sedimentation, the threat of destruction is multiplied a thousandfold by man's foolishness. Already, thousands of square kilometres of coral reef have fallen victim to blasting, construction work, bomb tests, factories, desalination plants and thermal power stations, pollution by sewage, toxic waste and oil, and have been stripped bare by the excessive zeal of collectors and hunters. Although these actions are undeniably a serious indictment of mankind who alone bears responsibility for the destruction, the damage appears so far to be "only" local. But if pollution of the oceans continues, the day is not far distant when life there will collapse. The sensitive coral reefs will be among the first major casualties. If we want to preserve the coral reefs, we must prevent both further pollution of the oceans of the world, and local destruction of the reefs. This book was written to widen the understanding of one of the most beautiful and richest biological communities on earth. May it in some measure help to rescue and preserve the coral reefs.

Classification and distribution

of recent Stony Corals (Scleractinia)

Consecutive numbers indicate genera, measurements in metres indicate bathymetric distribution as far as it is known, all remaining abbreviations indicate geographical distribution: Atl = Atlantic, AuI = Australasian intercontinental oceans, Ind = Indian Ocean, Indopac = Indo-Pacific, Car = Caribbean, American intercontinental oceans, Med = Mediterranean, Pac = Pacific, RS = Red Sea, cos = cosmopolitan, N = north, S = south, E = east, W = west

Order:

Scleractinia BOURNE 1900

Suborder:
Astrocoeniina VAUGHAN & WELLS 1943

Family: Astrocoeniidae KOBY 1890
1. *Actinastrea* D'ORBIGNY 1849; Cos
2. *Stephanocoenia* EDWARDS & HAIME 1848; Car, 0–95 m
3. *Stylocoeniella* YABE & SUGIYAMA 1935; Indopac, 0–58 m
Family: Thamnasteriidae VAUGHAN & WELLS 1943
4. *Psammocora* DANA 1846; Indopac, 0–140 m
Family: Pocilloporidae GRAY 1842
5. *Madracis* EDWARDS & HAIME 1849; Car-Indopac, 1–708 m
6. *Palauastrea* YABE & SUGIYAMA 1941; WPac, 1–10 m
7. *Pocillopora* LAMARCK 1815; Indopac, 0–78 m
8. *Seriatopora* LAMARCK 1816; Indopac, 0–55 m
9. *Stylophora* SCHWEIGGER 1816; Indopac, 0–70 m
Family: Acroporidae VERRILL 1902
10. *Acropora* OKEN 1815; Car-Indopac, 0–50 m
11. *Anacropora* RIDLEY 1884; Indopac
12. *Astreopora* BLAINVILLE 1830; Indopac, 1–20 m
13. *Montipora* QUOY & GAIMARD 1830; Indopac, 0–266 m

Suborder:
Fungiina VERRILL 1865

Superfamily:
Agariciicae GRAY 1847
Family: Agariciidae GRAY 1847
14. *Agaricia* LAMARCK 1801; Car, 0–86 m
15. *Coeloseris* VAUGHAN 1918; Indopac, 0–22 m
16. *Domoseris* QUELCH 1886; Indopac
17. *Gardineroseris* SCHEER & PILLAI 1974; Indopac, 0–40 m
18. *Leptoseris* EDWARDS & HAIME 1849; Car-Indopac, 0–165 m
19. *Pachyseris* EDWARDS & HAIME 1849; Indopac, 3–203 m
20. *Pavona* LAMARCK 1801; Indopac, 0–78 m
Family: Siderastreidae VAUGHAN & WELLS 1943
21. *Anomastraea* MARENZELLER 1901; Indopac
22. *Coscinaraea* EDWARDS & HAIME 1848; Indopac, 1–270 m
23. *Craterastrea* HEAD 1982; RS
24. *Horastrea* PICHON 1971; WInd, 0–40 m
25. *Pseudosiderastrea* YABE & SUGIYAMA 1935; WPac, 1–15 m
26. *Siderastrea* BLAINVILLE 1830; WInd-Car, 0–70 m
Superfamily:
Fungiicae DANA 1846
Family: Fungiidae DANA 1846
27. *Ctenactis* AGASSIZ 1860; Indopac, 0–25 m
28. *Cycloseris* EDWARDS & HAIME 1849; Indopac, 2–463 m
29. *Diaseris* EDWARDS & HAIME 1849; Ind-WPac, 5–30 m
30. *Fungia* LAMARCK 1801; Indopac, 0–60 m
31. *Fungiacyathus* SARS 1872; Cos, 283–5,872 m
32. *Halomitra* DANA 1846; Indopac, 1–35 m
33. *Heliofungia* WELLS 1966; Ind-WPac, 1–30 m
34. *Herpetoglossa* WELLS 1966; Indopac, 0–25 m
35. *Herpolitha* ESCHSCHOLTZ 1825; Indopac, 0–40 m
36. *Lithophyllon* REHBERG 1892; Indopac, 0–22 m
37. *Parahalomitra* WELLS 1937; Indopac, 0–20 m
38. *Podabacia* EDWARDS & HAIME 1849; Indopac, 0–50 m
39. *Polyphyllia* QUOY & GAIMARD 1833; Pac, 2–15 m
40. *Sandalolitha* QUELCH 1884; Indopac, 0–27 m

41. *Zoopilus* DANA 1846; Pac
Family: Micrabaciidae VAUGHAN 1905
42. *Leptopenus* MOSELEY 1881; SAtl-Car-SIndopac, 2,000–3,566 m
43. *Letepsammia* YABE & EGUCHI 1932; SPac, 143–192 m
44. *Micrabacia* EDWARDS & HAIME 1849; Cos
45. *Stephanophyllia* MICHELIN 1841; WPac, 10–283 m
Superfamily:
Poriticae GRAY 1842
Family: Poritidae GRAY 1842
46. *Alveopora* BLAINVILLE 1830; Indopac, 0–70 m
47. *Goniopora* BLAINVILLE 1830; Indopac, 1–35 m
48. *Napopora* QUELCH 1886; Pac, 1–40 m
49. *Porites* LINK 1807; Cos, 0–72 m
50. *Synaraea* VERRILL 1864; Indopac, 0–70 m

Suborder:
Faviina VAUGHAN & WELLS 1943

Superfamily:
Faviicae GREGORY 1900
Family: Faviidae GREGORY 1900
Subfamily: Faviinae GREGORY 1900
51. *Astraeosmilia* ORTMANN 1892; WInd
52. *Australogyra* VERON & PICHON 1982; WPac, 3–12 m
53. *Barabattoia* YABE & SUGIYAMA 1941; Pac
54. *Bikinastrea* WELLS 1954; Pac
55. *Caulastraea* DANA 1846; Indopac, 0–35 m
56. *Colpophyllia* EDWARDS & HAIME 1848; Car, 0–55 m
57. *Diploria* EDWARDS & HAIME 1848; Car, 0–43 m
58. *Erythrastrea* PICHON, SCHEER & PILLAI 1981; RS
59. *Favia* OKEN 1815; Cos, 0–100 m
60. *Favites* LINK 1807; Indopac, 0–50 m
61. *Goniastrea* EDWARDS & HAIME 1848; Indopac, 0–80 m
62. *Hydnophora* FISCHER 1807; Indopac, 0–55 m
63. *Leptoria* EDWARDS & HAIME 1848; Indopac, 0–35 m
64. *Manicina* EHRENBERG 1834; Car, 0–65 m
65. *Montigyra* MATTHAI 1928; Ind

66. *Oulophyllia* EDWARDS & HAIME 1848; Indopac, 2–57 m
67. *Platygyra* EHRENBERG 1834; Indopac, 0–40 m

Subfamily: Montastreinae VAUGHAN & WELLS 1943

68. *Cladocora* EHRENBERG 1834; Med-Atl-Car-EPac, 0–600 m
69. *Cyphastrea* EDWARDS & HAIME 1848; Indopac, 0–81 m
70. *Diploastrea* MATTHAI 1914; Indopac, 0–25 m
71. *Echinopora* LAMARCK 1816; Indopac, 0–50 m
72. *Leptastrea* EDWARDS & HAIME 1848; Indopac, 0–76 m
73. *Montastrea* BLAINVILLE 1830; Car-Indopac, 0–95 m
74. *Oulastrea* EDWARDS & HAIME 1848; Indopac, 10 m
75. *Plesiastrea* EDWARDS & HAIME 1848; Indopac
76. *Solenastrea* EDWARDS & HAIME 1848; Car-EPac, 0–23 m

Family: Trachyphylliidae VERRILL 1901

77. *Wellsophyllia* PICHON 1980; WPac, 6–8 m
78. *Moseleya* QUELCH 1884; WPac, 0–25 m
79. *Trachyphyllia* EDWARDS & HAIME 1848; Ind-WPac, 0–46 m

Family: Rhizangiidae D'ORBIGNY 1851

80. *Astrangia* EDWARDS & HAIME 1848; NAtl-Med-Car-WPac, 0–150 m
81. *Cladangia* EDWARDS & HAIME 1851; Ind
82. *Colangia* POURTALÈS 1871; Car, 3–95 m
83. *Culicia* DANA 1846; Indopac, 10–100 m
84. *Oulangia* EDWARDS & HAIME 1848; Indopac, 0–54 m
85. *Phyllangia* EDWARDS & HAIME 1848; Med-Atl-Car-EPac, 0–793 m

Family: Oculinidae GRAY 1847
Subfamily: Oculininae GRAY 1847

86. *Archohelia* VAUGHAN 1919; WPac, 3–4 m
87. *Bathelia* MOSELEY 1881; SAtl
88. *Cyathelia* EDWARDS & HAIME 1849; Indopac, 15–812 m
89. *Madrepora* LINNÉ 1758; Cos, 53–2,700 m
90. *Neohelia* MOSELEY 1881; Pac, 91–115 m
91. *Oculina* LAMARCK 1816; Atl-Med-Car-Pac, 0–260 m
92. *Slerhelia* EDWARDS & HAIME 1850; SAtl-Indopac

Subfamily: Galaxeinae VAUGHAN & WELLS 1943

93. *Acrhelia* EDWARDS & HAIME 1849; WPac, 0–25 m
94. *Galaxea* OKEN 1815; Indopac, 0–60 m
95. *Simplastrea* UMBGROVE 1939; AuI

Family: Meandrinidae GRAY 1847
Subfamily: Meandrininae GRAY 1847

96. *Ctenella* MATTHAI 1928; Ind
97. *Goreaugyra* WELLS 1973; Car, 26–30 m
98. *Meandrina* LAMARCK 1801; WAtl-Car, 0–80 m

Subfamily: Dichocoeniinae VAUGHAN & WELLS 1943

99. *Dendrogyra* EHRENBERG 1834; Car, 2–20 m
100. *Dichocoenia* EDWARDS & HAIME 1848; Car, 0–47 m

Family: Merulinidae VERRILL 1866

101. *Boninastraea* YABE & SUGIYAMA 1935; Pac
102. *Clavarina* VERRILL 1864; WPac, 0–35 m
103. *Merulina* EHRENBERG 1834; Indopac, 0–35 m
104. *Scapophyllia* EDWARDS & HAIME 1848; EInd-WPac, 1–25 m

Family: Mussidae ORTMANN 1890

105. *Acanthastrea* EDWARDS & HAIME 1848; Indopac, 0–40 m
106. *Acanthophyllia* WELLS 1937; Indopac, 16–75 m
107. *Blastomussa* WELLS 1961; Indopac, 0–50 m
108. *Cynarina* BRUEGGEMANN 1877; Indopac, 0–75 m
109. *Homophyllia* BRUEGGEMANN 1877; WPac, 0–5 m
110. *Isophyllastraea* MATTHAI 1928; Car, 1–20 m
111. *Isophyllia* EDWARDS & HAIME 1851; Car, 1–10 m
112. *Lobophyllia* BLAINVILLE 1830; Indopac, 0–55 m
113. *Mussa* OKEN 1815; Car, 1–59 m
114. *Mussismilia* ORTMANN 1890; WAtl
115. *Mycetophyllia* EDWARDS & HAIME 1848; Car, 1–76 m
116. *Scolymia* EDWARDS & HAIME 1852; Car-Indopac, 0–140 m
117. *Symphyllia* EDWARDS & HAIME 1848; Indopac, 0–55 m

Family: Pectiniidae VAUGHAN & WELLS 1943

118. *Echinophyllia* KLUNZINGER 1879; Indopac, 0–180 m
119. *Mycedium* OKEN 1815; Indopac, 0–70 m
120. *Oxypora* SAVILLE-KENT 1871; Pac, 0–35 m
121. *Pectinia* OKEN 1815; Indopac, 0–40 m
122. *Physophyllia* DUNCAN 1884; Pac, 35 m

Family: Anthemiphylliidae VAUGHAN 1907

123. *Anthemiphyllia* POURTALÈS 1878; Car-Indopac, 50–732 m
124. *Bathytrochus* GRAVIER 1915; Atl, 4,023 m

Suborder:
Caryophylliina VAUGHAN & WELLS 1943

Superfamily:
Caryophylliicae GRAY 1847
Family: Caryophylliidae GRAY 1847
Subfamily: Caryophylliinae GRAY 1847

125. *Aulocyathus* MARENZELLER 1904; WInd, 366–1,350 m
126. *Bathycyathus* EDWARDS & HAIME 1848; NAtl-Car-EPac, 55–165 m
127. *Caryophyllia* LAMARCK 1801; Cos, 0–3,100 m
128. *Ceratotrochus* EDWARDS & HAIME 1848; Cos, 7–732 m

129. *Concentrotheca* CAIRNS 1979; NAtl-Car, 183–800 m
130. *Coenocyathus* EDWARDS & HAIME 1848; NAtl-Car-EPac, 0–732 m
131. *Cyathoceras* MOSELEY 1881; Car-SAtl-Indopac, 220–1,372 m
132. *Dasmosmilia* POURTALÈS 1880; Atl-Car-RS, 48–600 m
133. *Deltocyathoides* YABE & EGUCHI 1932; WPac, 91–355 m
134. *Deltocyathus* EDWARDS & HAIME 1848; Cos, 13–4,480 m
135. *Fragilocyathus* YABE & EGUCHI 1932; WPac, 84–289 m
136. *Heterocyathus* EDWARDS & HAIME 1848; Indopac, 11–658 m
137. *Lachmaeotrochus* ALCOCK 1902; Aul, 366–530 m
138. *Labyrinthocyathus* CAIRNS 1979; Car, 385–810 m
139. *Lophosmilia* EDWARDS & HAIME 1848; Atl-Pac, 58–366 m
140. *Nomlandia* DURHAM & BARNARD 1952; EPac, 80–90 m
141. *Oxysmilia* DUCHASSAING 1870; Car, 46–640 m
142. *Paracyathus* EDWARDS & HAIME 1848; Cos, 0–2,000 m
143. *Polycyathus* DUNCAN 1876; Atl-WInd, 20–73 m
144. *Stephanocyathus* SEGUENZA 1864; Cos, 141–2,235 m
145. *Tethocyathus* KÜHN 1933; Car-Ind-Aul, 13–888 m
146. *Trochocyathus* EDWARDS & HAIME 1848; 32–2,510 m
147. *Vaughanella* GRAVIER 1915; NAtl, 545–3,013 m

Subfamily: Turbinoliinae EDWARDS & HAIME 1848

148. *Conocyathus* D'ORBIGNY 1849; Indopac, 59–73 m
149. *Cylindrophyllia* YABE & EGUCHI 1937; Pac, 260 m
150. *Dunocyathus* TENISON-WOODS 1878; WPac, 82–457 m
151. *Holcotrochus* DENNANT 1902; WPac, 10–41 m
152. *Idiotrochus* WELLS 1936; WPac
153. *Kionotrochus* DENNANT 1906; Pac, 48–470 m
154. *Notocyathus* TENISON-WOODS 1880; NAtl-WPac, 55–835 m
155. *Peponocyathus* GRAVIER 1915; Atl-Car-Pac, 110–1,097 m
156. *Platytrochus* EDWARDS & HAIME 1848; Pac, 27–130 m
157. *Sphenotrochus* EDWARDS & HAIME 1848; Cos, 15–274 m
158. *Trematotrochus* TENISON-WOODS 1879; Car-Pac, 27–576 m
159. *Tropidocyathus* EDWARDS & HAIME 1848; Indopac, 11–480 m
160. *Turbinolia* LAMARCK 1816; Atl-Car, 183–567 m

Subfamily: Desmophyllinae VAUGHAN & WELLS 1943

161. *Dactylotrochus* WELLS 1954; Pac, 75–135 m
162. *Desmophyllum* EHRENBERG 1834; Cos, 0–3,100 m
163. *Gemmulatrochus* DUNCAN 1878; NEAtl-Med
164. *Lophelia* EDWARDS & HAIME 1849; Atl-Indopac, 60–2,875 m
165. *Thalamophyllia* DUCHASSAING 1870; Atl-Med-Car, 18–1,317 m
Subfamily: Parasmiliinae VAUGHAN & WELLS 1943
166. *Anomocora* STUDER 1878; Atl-Car-Aul, 50–576 m
167. *Asterosmilia* DUNCAN 1867; NAtl-Car, 32–1,435 m
168. *Coenosmilia* POURTALÈS 1874; NAtl-Car, 109–2,237 m
169. *Dendrosmilia* EDWARDS & HAIME 1848; NPac, 178 m
170. *Goniocorella* YABE & EGUCHI 1932; Ind-WPac, 55–690 m
171. *Parasmilia* EDWARDS & HAIME 1848; Cos, 313–366 m
172. *Phacelocyathus* CAIRNS 1979; Car, 22–560 m
173. *Pourtalosmilia* DUNCAN 1884; NAtl-Med, 110–300 m
174. *Rhizosmilia* CAIRNS 1978; Car, 123–355 m
175. *Solenosmilia* DUNCAN 1873; Atl-Car-Indopac, 220–3,465 m
Subfamily: Eusmiliinae EDWARDS & HAIME 1857
176. *Catalaphyllia* WELLS 1971; Ind-WPac, 2–50 m
177. *Euphyllia* DANA 1846; Indopac, 0–35 m
178. *Eusmilia* EDWARDS & HAIME 1848; Car, 0–65 m

179. *Gyrosmilia* EDWARDS & HAIME 1851; RS
180. *Nemenzophyllia* HODGSON & ROSS 1981; WPac, 15 m
181. *Physogyra* QUELCH 1884; Indopac, 0–25 m
182. *Plerogyra* EDWARDS & HAIME 1848; Indopac, 0–45 m
Superfamily:
Flabellicae BOURNE 1905
Family: Flabelliidae BOURNE 1905
183. *Flabellum* LESSON 1831; Cos, 3–3,183 m
184. *Gardineria* VAUGHAN 1907; SAtl-Car-Pac, 10–700 m
185. *Javania* DUNCAN 1876; NAtl-Med-Car, 86–2,165 m
186. *Monomyces* EHRENBERG 1834; NAtl-Med-Indopac, 27–1,047 m
187. *Placotrochides* ALCOCK 1902; NAtl-Car, 497–1,300 m
188. *Placotrochus* EDWARDS & HAIME 1848; NAtl-Car-Indopac, 265–1,300 m
189. *Polymyces* CAIRNS 1979; NWAtl, 75–798 m
Family: Guyniidae HICKSON 1910
190. *Guynia* DUNCAN 1873; Med-Car-WInd-WPac, 3–658 m
191. *Pourtalocyathus* CAIRNS 1979; Car, 349–1,200 m
192. *Schizocyathus* POURTALÈS 1874; Atl-Car, 88–1,445 m
193. *Stenocyathus* POURTALÈS 1871; Atl-Med-Car-WPac, 80–1,229 m

Suborder:
Dendrophylliina VAUGHAN & WELLS 1943

Family: Dendrophylliidae GRAY 1847
194. *Astroides* QUOY & GAIMARD 1827; Med, 0–85 m
195. *Balanophyllia* S. WOOD 1884; Cos, 0–1,150 m
196. *Bathypsammia* MARENZELLER 1906; WAtl-Car, 210–1,079 m
197. *Cladopsammia* LACAZE-DUTHIERS 1897; Med, 12–50 m
198. *Coenopsammia* EDWARDS & HAIME 1848; Indopac, 0–200 m
199. *Dendrophyllia* BLAINVILLE 1830; Cos, 0–1,372 m
200. *Duncanopsammia* WELLS 1936; WPac, 1–25 m
201. *Enallopsammia* MICHELOTTI 1871; NAtl-Car-Indopac, 229–3,389 m
202. *Endopachys* LONSDALE 1845; Indopac, 37–604 m
203. *Endopsammia* EDWARDS & HAIME 1848; SAtl-Med-Indopac, 2–91 m
204. *Heteropsammia* EDWARDS & HAIME 1848; Indopac, 11–192 m
205. *Leptopsammia* EDWARDS & HAIME 1848; NAtl-Med, 1–1,385 m
206. *Notophyllia* DENNANT 1899; WPac, 37–457 m
207. *Psammoseris* EDWARDS & HAIME 1851; AuI, 11–44 m
208. *Rhizopsammia* VERRILL 1869; NAtl-Indopac, 0–200 m
209. *Thecopsammia* POURTALÈS 1868; NAtl-Car-Pac, 0–1,097 m
210. *Trochopsammia* POURTALÈS 1878; Car, 403–1,490 m
211. *Tubastraea* LESSON 1834; Cos, 0–1,463 m
212. *Turbinaria* OKEN 1815; Indopac, 0–35 m

Further Reading on the Determination of Stony Corals

CAIRNS, S. D.: "The deep-water scleractinia of the Caribbean Sea and adjacent waters", in: *Stud. Fauna Curaçao 80.* 341 pp., 40 plates, 1979.

CAIRNS, S. D.: "Antarctic and subantarctic Scleractinia", in: *Biology of the Antarctic Seas 11, 1.* 74 pp., 1982.

CHEVALIER, J.-P.: "Les Scléractiniaires de la Mélanésie française, Pt. 1. Expéd. Franç. Récifs Corall. Nouvelle-Calédonien", in: *Fond. Singer-Polignac Paris.* 307 pp., 182 ill., 38 plates, 1971.

CHEVALIER, J.-P.: "Les Scléractiniaires de la Mélanésie française, Pt. 2. Expéd. Franç. Récifs Corall. Nouvelle-Calédonie", in: *Fond Singer-Polignac Paris.* 407 pp., 250 ill., 42 plates, 1975.

DEAS, W., and S. DOMM: "Corals of the Great Barrier Reef", in: *Ure Smith Ltd. Sydney.* 127 pp., 189 ill., 1976.

DITLEV, H.: "A Field-guide to the reef-building Corals of the Indo-Pacific", in: *Bachhuys Publ. Rotterdam and Scandinavian Science Press Klempenborg.* 291 pp., 390 ill., 1980.

DUARTE BELLO, P. P.: "Corales de los arrecifes Cubanos", in: *Acuario Nacional La Habana, Ser. educ. 2.* 85 pp., 74 ill., 1961.

EGUCHI, M.: "The hydrocorals and scleractinian corals of Sagami Bay", in: *Maruzen Tokyo.* 382 pp., 100 plates, 1968.

FAULKNER, D., and R. CHESHER: "Living Corals", in: *Clarkson N. Potter Inc. Publ.* New York, 310 pp., 194 plates, 1979.

PILLAI, C. S. G., and G. SCHEER: "Report on the Stony Corals from the Maldive Archipelago", in: *Zoologica 126.* 83 pp., 32 plates, 1976.

ROOS, P. J.: "The shallow-water stony corals of the Netherlands Antilles", in: *Stud. Curaçao 37.* 108 pp., 53 plates, 1971.

SCHEER, G., and C. S. G. PILLAI: "Report on the Scleractinia from the Nicobar Islands", in: *Zoologica 122.* 75 pp., 33 plates, 1974.

SCHEER, G., and C. S. G. PILLAI: "Report on the stony corals from the Red Sea", in: *Zoologica 133.* 198 pp., 41 plates, 1983.

SMITH, F. G. W.: "Atlantic reef corals", in: *Univ. Miami Press,* Coral Gables Florida. 164 pp., 48 plates, 1971.

SQUIRES, D. F., and I. W. KEYES: "The marine fauna of New Zealand: Scleractinian corals", in: *Mem. N. Z. Oceanogr. Inst. 43.* 46 pp., 1967.

VERON, J. E. N., and M. PICHON: "Scleractinia of Eastern Australia I, Families Thamnasteriidae, Astrocoeniidae, Pocilloporidae", in: *Austral. Inst. Mar. Sci. Monogr. Ser. 1.* 86 pp., 166 ill., 1976.

VERON, J. E. N., and M. PICHON: Scleractinia of Eastern Australia III, Families Agariciidae, Siderastreidae, Fungiidae, Oculinidae, Merulinidae, Mussidae, Pectiniidae, Caryophylliidae, Dendrophylliidae", in: *Austral. Inst. Mar. Sci. Monogr. Ser. 4.* 460 pp., 857 ill., 1979.

VERON, J. E. N., and M. PICHON: "Scleractinia of Eastern Australia IV, Family Poritidae", in: *Austral. Inst. Mar. Sci. Monogr. Ser. 5.* 159 pp., 346 ill., 1982.

VERON, J. E. N., M. PICHON and M. WIJSMAN-BEST: "Scleractinia of Eastern Australia II, Families Faviidae, Trachyphylliidae", in: *Austral. Inst. Mar. Sci. Monogr. Ser. 3.* 233 pp., 477 ill., 1977.

WALLACE, C. C.: "The coral genus Acropora (Scleractinia: Astrocoeniina: Acroporidae) in the Central and Southern Great Barrier Reef Province", in: *Mem. Queensland Mus. 18.* pp. 273-319, 60 plates, 1978.

WELLS, J. W.: "Recent corals of the Marshall Islands", in: *U.S. Geol. Survey Prof. Pap. 260-1.* pp. 385-486, 93 plates, 1954.

WELLS, J. W.: "Scleractinia", in: *Treatise on Invertebrate Paleontology, Pt. F.* edit. by R. C. Moore, Geol. Soc. Amer., Univ. Kansas Press, pp. 328-444, 117 ill., 1963.

ZIBROWIUS, H.: "Les Scléractiniaires de la Méditerranée et de l'Atlantique nord-oriental", in: *Mém. Inst. Océanogr. Monaco 11.* 284 pp., 107 plates, 1980.

ZLATARSKI, V. N., and N. MARTINEZ ESTALELLA: *Les Scléractiniaires de Cuba avec des données sur les organismes associés.* Edit. Acad. Bulgare des Sciences Sofia, 472 pp., 160 plates, 138 ill., 1982.

Bibliography

ADEY, W. H., and I. G. MACINTYRE: "Crustose coralline algae: a re-evaluation in the geological sciences", in: *Geol. Soc. Amer. Bull. 84,* pp. 883-904, 1973.

ADEY, W. H., and J. M. VASSAR: "Colonization succession and growth rates of tropical crustose coralline algae (Rhodophyta, Cryptonemiales)", in: *Phycologia 14,* pp. 55-69, 1975.

ALLEN, G. R.: *Falter- und Kaiserfische.* Vol. 2. Melle, 1979.

BAK, R. P. M., and G. VAN EYS: "Predation of the sea urchin *Diadema antillarum* Philippi on living coral", in: *Oecologia 20,* pp. 111-115, 1975.

BARLOW, G. W.: "On the sociobiology of some hermaphroditic serranid fishes, the hamlets, in Puerto Rico", in: *Mar. Biol. 33,* pp. 295-300, 1975.

BEMERT, G., and R. ORMOND: *Red Sea coral reefs.* London, Boston, 1981.

BENNETT, I.: *The Great Barrier Reef.* Melbourne, 1971.

BOUILLON, J., and N. HOUVENAGHEL-CREVECOEUR: "Etude monographique du genre *Heliopora* de Blainville", in: *Ann. Mus. Roy. Afr. Centr. Tervuren. Ser. 8. Zool., 178,* pp. 1-83, 1970.

BROMLEY, R. G.: "Bioerosion of Bermuda Reefs", in: *Palaeogr. Palaeoclimat. Palaeoecol. 23,* pp. 169-197, 1978.

CASTRO, P.: "Movements between coral colonies in *Trapezia ferruginea* (Crustacea, Brachyura), an obligate symbiont of scleractinian corals", in: *Mar. Biol. 46,* pp. 237-245, 1978.

COLES, S. L., and P. L. JOKIEL: "Effects of temperature on photosynthesis and respiration in hermatypic coral", in: *Mar. Biol. 43,* pp. 209-216, 1977.

DALY, R. A.: "The glacial-control theory of coral reefs", in: *Proc. Amer. Acad. Arts Sci. Boston 51,* pp. 155-251, 1915.

DANA, J. D.: *Corals and Coral Islands.* London, 1875.

DARWIN, C.: "On certain areas of elevation and subsidence in the Pacific and Indian Oceans, as deduced from the study of coral formations", in: *Proc. Geol. Soc. London 2,* pp. 552-554, 1837.

DARWIN, C.: *Über den Bau und die Verbreitung der Corallenriffe.* Stuttgart, 1899.

DAWSON, E. Y.: "Changes in Palmyra Atoll and its vegetation through the activities of man 1913-1958", in: *Pacif. Nat. 1,* pp. 1-51, 1959.

DONALDSON, L. R.: "Radiobiological studies at the Eniwetok test site and adjacent areas of the western Pacific", in: *Trans. 2. Seminar of Biol.*

Probl. in Water Pollution. Washington 1–7, 1960.

DUSTAN, P.: "Distribution of zooxanthellae and photosynthetic chloroplast pigments of the reef-building coral *Montastrea annularis* Ellis and Solander in relation to depth on a West Indian coral reef", in: *Bull. Mar. Sci. 29,* pp. 79–95, 1979.

EIBL-EIBESFELDT, I.: *Galapagos.* Munich, no date.

EIBL-EIBESFELDT, I.: *Im Reich der tausend Atolle.* Munich, 1964.

FISHELSON, L.: "Ecology of the northern Red Sea crinoids and their epi- and endozoic fauna", in: *Mar. Biol. 26,* pp. 183–192, 1974.

FISHELSON, L., POPPER, D., and A. AVIDOR: "Biosociology and ecology of pomacentrid fishes around the Sinai peninsula (northern Red Sea)", in: *J. Fish. Biol. 6,* pp. 119–133, 1974.

FORSTER, G.: *Reise um die Welt.* Leipzig, no date.

FRANZISKET, L.: "Nitrate uptake by reef corals", in: *Int. Revue ges. Hydrobiol. 59,* pp. 1–7, 1974.

FRICKE, H. W.: *Korallenmeer.* Stuttgart, 1972.

FRICKE, H. W.: "Öko-Ethologie des monogamen Anemonenfisches *Amphiprion bicinctus*", in: *Tierpsychol. 36,* pp. 429–513, 1974.

FRICKE, H. W.: "Sozialstruktur und ökologische Spezialisierung von verwandten Fischen (Pomacentridae)", in: *Tierpsychol. 39,* pp. 492–520, 1975.

FRICKE, H. W.: *Bericht aus dem Riff.* Munich, Zurich, 1976.

FRICKE, H. W., and E. VARESCHI: "A scleractinian coral *(Plerogyra sinuosa)* with 'photosynthetic organs'", in: *Mar. Ecol. Prog. Ser. 7,* pp. 273–278, 1982.

FROST, S. H., WEISS, M. P., and J. B. SAUNDERS (ed.): *Reefs and related carbonate-ecology and sedimentology.* Tulsa, 1977.

GARDINER, J. S.: *Coral reefs and atolls.* London, 1931.

GARDINER, J. S.: "Photosynthesis and solution in formation of coral reefs", in: *Nature 127,* pp. 857–858, 1931.

GOHAR, H. A. F., and G. N. SOLIMAN: "On the biology of three coralliophilids boring in living corals", in: *Publ. Mar. Biol. Stat. Ghardaqa 12,* pp. 99–126, 1963.

GOREAU, T. F.: "The ecology of Jamaican reefs. I. Species composition and zonation", in: *Ecology 40,* pp. 67–90, 1959.

GOREAU, T. F.: "On the relation of calcification to primary productivity in reef building organisms", in: H. M. LENHOFF, and W. F. LOOMIS (ed.): *The Biology of Hydra and some other coelenterates,* pp. 269–285, 1961.

GOREAU, T. F., GOREAU, N. I., and C. M. YONGE: "Reef corals: autothrophs or heterotrophs?", in: *Biol. Bull. 141,* pp. 247–260, 1971.

GRASSHOFF, M.: "Polypen und Kolonien der Blumentiere (Anthozoa)", in: *Nat. Mus. 111,* pp. 1–8, 29–45, 134–150, 1981.

GRAUS, R. R., and I. G. MACINTYRE: "Light control of growth form in colonial reef corals: computer simulation", in: *Science 193,* pp. 895–897, 1976.

HADFIELD, M. G.: "Molluscs associated with living tropical corals", in: *Micronesica 12,* pp. 133–148, 1976.

HARTMAN, W. D., and T. F. GOREAU: "Jamaican coralline sponges: Their morphology, ecology and fossil relatives", in: *Symp. Zool. Soc. London 25,* pp. 205–243, 1970.

HASS, H.: "Central subsidence. A new theory of atoll formation", in: *Atoll Res. Bull. No. 91,* 4 pp., 1962.

HOFFMEISTER, J. E.: *Land from the sea.* Coral Gables, 1976.

HOLZBERG, S.: "Beobachtung einer Putzsymbiose zwischen der Garnele *Leandrites cyrtirhynchus* und Riffbarschen", in: *Helgoländer wiss. Meeresunters. 22,* pp. 362–365, 1971.

HUMBOLDT, A. V.: *Ansichten der Natur.* Stuttgart, 1849.

HUMES, A. G.: "Coral-inhabiting copepods from the Moluccas with a synopsis of cyclopoids, associated with scleractinian corals", in: *Cah. Biol. Mar. 20,* pp. 77–107, 1979.

JOHANNES, R. E.: "Pollution and degradation of coral reef communities", in: E. J. F. WOOD, and R. E. JOHANNES (ed.): "Tropical Pollution", in: *Elsevier Oceanogr. Ser. 12,* pp. 13–51, 1975.

JONES, O. A., and R. ENDEAN (ed.): *Biology and geology of coral reefs.* 4 vols., New York, London, 1973–1977.

KAWAGUTI, S.: "On the physiology of reef corals. IV. The growth of *Goniastrea aspera* measured from numerical and areal increase of calyces", in: *Palao Trop. Biol. Stat. Stud. Tokyo 2,* pp. 309–317, 1941.

KESTERMAN, F., and E. L. TOWLE: "Caribbean weighs impact of stepped up oil industry activity", in: *J. Marit. Law Commer. 4,* pp. 517–523, 1973.

KOHN, A. J.: "Diversity, utilization of resources and adaptive radiation in shallow-water marine invertebrates of tropical oceanic islands", in: *Limnol. Oceanogr. 16,* pp. 332–348, 1971.

KÜHLMANN, D. H. H.: "Zur Anwendung der Direktbeobachtungen in der Hydrobiologie mittels autonomer Tauchgeräte", in: *Hidrobiol. (Bucureşti) 3,* pp. 207–212, 1961.

KÜHLMANN, D. H. H.: "Die Korallenriffe Kubas. I. Genese und Evolution", in: *Int. Revue ges. Hydrobiol. 55,* pp. 729–756, 1970.

KÜHLMANN, D. H. H.: "Studien über physikalische und chemische Faktoren in kubanischen Riffgebieten", in: *Acta Hydrophysica 15,* pp. 105–152, 1970.

KÜHLMANN, D. H. H.: "Zur Methodik der Korallenriffuntersuchungen", in: *Wiss. Z. Humboldt-Univ., Math.-Nat. R. 20,* pp. 697–705, 1971.

KÜHLMANN, D. H. H.: "Die nomographische Ermittlung von Sauerstoff-Sättigungskonzentration, Sauerstoff-Sättigungsindex und respiratorischem Wert in Süß-, Brack- und Meerwasser", in: *Acta Hydrophysica 16,* pp. 27–35, 1971.

KÜHLMANN, D. H. H.: "Die Korallenriffe Kubas. II. Zur Ökologie der Bankriffe und ihrer Korallen", in: *Int. Revue ges. Hydrobiol. 56,* pp. 145–199, 1971.

KÜHLMANN, D. H. H.: "Über Korallen und Korallenriffe des westlichen Indischen Ozeans und der karibischen Region — eine vergleichende Studie", in: *J. Mar. Biol. Ass. India 14,* pp. 512–523, 1972.

KÜHLMANN, D. H. H.: "Die Korallenriffe Kubas. III. Riegelriff und Korallenterasse, zwei verwandte Erscheinungen des Bankriffs", in: *Int. Revue ges. Hydrobiol. 59,* pp. 305–325, 1974.

KÜHLMANN, D. H. H.: "The coral reefs of Cuba", in: *Proc. Second Int. Coral Reef Symp. 2,* Brisbane, pp. 69–83, 1974.

KÜHLMANN, D. H. H.: "Notes on the influence of particulate organic matter (POM) on light conditions at a coral reef of the Australian Great Barrier Reef", in: *Int. Revue ges. Hydrobiol. 60,* pp. 195–198, 1975.

KÜHLMANN, D. H. H.: "Charakterisierung der Korallenriffe vor Veracruz/Mexico", in: *Int. Revue ges. Hydrobiol. 60,* pp. 495–521, 1975.

KÜHLMANN, D. H. H.: "Coral reefs—their importance and imperilment. UNESCO International postgraduate training course on ecosystem management", in: *Tech. Univ. Dresden 5, 4,* pp. 58–63, 1978.

KÜHLMANN, D. H. H.: "Coral associations and their value for palaeontological research", in: *Acta palaeont. Polon. 25,* pp. 459–466, 1980.

KÜHLMANN, D. H. H.: "Darwin's coral reef research—a review and tribute", in: *Mar. Ecol. 3,* pp. 193–212, 1982.

KÜHLMANN, D. H. H.: "Composition and ecology of deepwater coral associations", in: *Helgoländer wiss. Meeresunters. 36,* pp. 183–204, 1983.

KÜHLMANN, D. H. H., and H. KARST: "Freiwasserbeobachtungen zum Verhalten von Tobiasfischschwärmen (Ammodytidae) in der westlichen Ostsee", in: *Tierpsychol. 24,* pp. 282–297, 1967.

KÜKENTHAL, W.: "Octocorallia", in: W. KÜKENTHAL and T. KRUMBACH (ed.): *Handbuch der Zoologie.* Vol. I, pp. 690–769, Berlin, Leipzig, 1925.

LANG, J. C.: "Interspecific aggression by scleractinian corals", in: *Bull. Mar. Sci. 21,* pp. 952–959, 1971; *23,* pp. 260–279, 1973.

LAPORTE, L. F. (ed.): "Reefs in time and space", in: *Soc. Econ. Paleont. Mineral.,* Spec. Publ. No. 18. Tulsa, 1974.

LASSIG, B. R.: "Communication and coexistence in a coral community", in: *Mar. Biol. 42,* pp. 85–92, 1977.

LAUCKNER, G.: "Diseases of Cnidaria", in: O. KINNE (ed.): *Diseases of Marine Animals 1,* pp. 169–237. Chichester, New York, Brisbane, Toronto, 1980.

LEWIS, J. B.: "The effect of crude oil and an oil spill dispersant on reef corals", in: *Mar. Pollut. Bull. 2,* pp. 59–62, 1971.

LOW, R. M.: "Interspecific territoriality in a pomacentrid reef fish, *Pomacentrus flavicauda* Whitley", in: *Ecology 52,* pp. 648–654, 1971.

LORENZ, K.: "Nachwort", in: H. W. FRICKE: *Bericht aus dem Riff.* pp. 235–240. Munich, Zurich, 1976.

MARSDEN, J. R.: "The digestive tract of *Hermodice carunculata* (Pallas), Polychaeta: Amphinomidae", in: *Can. J. Zool. 41,* pp. 165–184, 1963.

MATHER, P., and I. BENNETT: *A coral reef handbook.* G. B. R. C. Handbook Ser. No. 1, 1978.

MAXWELL, W. G. H.: *Atlas of the Great Barrier Reef*. Amsterdam, 1968.

MILLIMAN, J. D.: *Marine Carbonates*. Berlin, Heidelberg, New York, 1974.

MOOSLEITNER, H.: "Korallenriff im Mittelmeer", in: *Delphin 21,* pp. 13–15, 1974.

MÜLLER, A. H.: *Lehrbuch der Paläozoologie 2: Invertebraten*. Part 1: *Protozoa-Mollusca 1.* Jena, 1968.

MYRBERG, A. A. jr., and R. E. THRESHER: "Interspecific aggression and its relevance to the concept of territoriality in reef fishes", in: *Amer. Zool. 14,* pp. 81–96, 1974.

NEUMANN, A. C.: "Observations on coastal erosion in Bermuda and measurements of the boring rate of the sponge, *Cliona lampa*", in: *Limnol. Oceanogr. 11,* pp. 92–108, 1966.

OGDEN, J. R., BROWN, R., and N. SALESKY: "Grazing by the echinoid *Diadema antillarum* Philippi: Formation of halos around West Indian patch reefs", in: *Science 182,* pp. 715–717, 1973.

ORMOND, R. F. G.: "Aggressive mimicry and other interspecific feeding associations among Red Sea coral reef predators", in: *J. Zool. Lond. 191,* pp. 247–262, 1980.

OTTER, G. W.: "Rock-destroying organisms in relation to coral reefs", in: *Sci. Rept. Great Barrier Reef Exped. 1928-1929, 1,* pp. 323–352, 1937.

PAX, F.: "Hexacorallia", in: W. KÜKENTHAL and T. KRUMBACH (ed.): *Handbuch der Zoologie.* Vol. I, pp. 770–901. Berlin, Leipzig, 1925.

PEARSON, R. G.: "Recovery and recolonization of coral reefs", in: *Mar. Ecol. Prog. Ser. 4,* pp. 105–122, 1981.

PLAYFORD, P. E.: "Australia's Stromatolite stronghold", in: *Nat. Hist. 89,* pp. 58–61, 1980.

POPPER, D., and L. FISHELSON: "Ecology and behavior of *Anthias squamipinnis* (Peters, 1955) (Anthiidae, Teleostei) in the coral habitat of Eilat (Red Sea)", in: *J. exp. Zool. 184,* pp. 409–424, 1973.

POTTS, D. C.: "Suppression of coral populations by filamentous algae within damselfish territories", in: *J. exp. mar. Biol. Ecol. 28,* pp. 297–312, 1977.

RANDALL, J. E., and P. GUÉZE: "The goatfish *Mulloidichthys mimicus* n. sp. (Pisces, Mullidae) from Oceania, a mimic of the snapper *Lutjanus kasmira* (Pisces, Lutjanidae)", in: *Bull. Mus. Nat. Hist., 4ᵉ Sér., 2,* pp. 603–609, 1980.

RANDALL, J. E., HEAD, S. M., and A. P. L. SANDERS: "Food habits of the giant humphead wrasse, *Cheilinus undulatus* (Labridae)", in: *Env. Biol. Fish. 3,* pp. 235–238, 1978.

RANDALL, J. E., and G. HELFMAN: "*Diproctacanthus xanthurus*, a cleaner wrasse from the Palau Islands, with notes on other cleaning fishes", in: *Trop. Fish Hobbyist 20,* pp. 87–95, 1972.

RANDALL, J. E., and V. G. SPRINGER: "The monotypic Indo-Pacific labrid fish genera *Labrichthys* and *Diproctanthus* with description of a new related genus, *Larabicus*", in: *Proc. Biol. Sci. Washington 86,* pp. 269–298, 1973.

REESE, E. S.: "Shell use: an adaptation for emigration from the sea by the coconut crab", in: *Science 161,* pp. 385–386, 1968.

RIEDL, R.: "Marine ecology—a century of changes", in: *Mar. Ecol 1,* pp. 3–46, 1980.

ROBERTSON, D. R.: "Field observations on the reproductive behaviour of a pomacentrid fish, *Acanthochromis polyacanthus*", in: *Tierpsychol. 32,* pp. 319–324, 1973.

RUDMAN, W. B.: "The anatomy and biology of alcyonarian-feeding aeolid opisthobranch molluscs and their development of symbiosis with zooxanthellae", in: *Zool. J. Linn. Soc. 72,* pp. 219–262, 1981.

RÜTZLER, K., and G. RIEGER: "Sponge burrowing: fine structure of *Cliona lampa* penetrating calcareous substrate", in: *Mar. Biol. 21,* pp. 144–162, 1973.

RUSSELL, B. C., ANDERSON, G. R., and F. H. TALBOT: "Seasonality and recruitment of coral reef fishes", in: *Aust. J. Mar. Freshwater Res. 28,* pp. 521–528, 1977.

RUSSELL, B. C., ALLEN, G. R., and H. R. LUBBOCK: "New cases of mimicry in marine fishes", in: *J. Zool. Lond. 180,* pp. 407–423, 1976.

SALE, P. F.: "Apparent effect of prior experience on a habitat preference exhibited by the reef fish, *Dascyllus aruanus* (Pisces: Pomacentridae)", in: *Anim. Behav. 19,* pp. 251–256, 1971.

SALVAT, B.: "Dégration des écosystèmes coralliens", in: *Courrier Nat. 30,* pp. 49–62, 1974.

SALVINI-PLAWEN, L. V.: "Cnidaria as food source for marine invertebrates", in: *Cah. Biol. Mar. 13,* pp. 385–400, 1972.

SCHEER, G.: "Über die Methodik der Untersuchung von Korallenriffen", in: *Morph. Ökol. Tiere 60,* pp. 105–114, 1967.

SCHLICHTER, D.: "Chemische Tarnung. Die stoffliche Grundlage der Anpassung von Anemonenfischen an Riffanemonen", in: *Mar. Biol. 12,* pp. 137–150, 1972.

SCHUHMACHER, H.: *Korallenriffe.* Munich, Berne, Vienna, 1976.

SCHUHMACHER, H.: "Experimentelle Untersuchungen zur Anpassung von Fungiiden (Scleractinia, Fungiidae) an unterschiedliche Sedimentations- und Bodenverhältnisse", in: *Int. Revue ges. Hydrobiol. 64,* pp. 207–243, 1979.

SHEPPARD, C. R. C.: "Coral populations on reef slopes and their major controls", in: *Mar. Biol. Prog. Ser. 7,* pp. 83–115, 1982.

SHINN, E. A.: "Spur and groove formation on the Florida reef tract", in: *J. Sed. Petrol. 33,* pp. 291–303, 1963.

SOROKIN, Y. L.: "Phytoplankton and planktonic bacteria in the ecosystem of coral reef", in: *Z. obsc. Biol. 40,* pp. 677–688, 1979.

STEENE, R. C.: *Falter- und Kaiserfische.* Vol. 1. Melle, 1977.

STODDART, D. R., and R. E. JOHANNES (ed.): *Coral Reefs: research methods.* UNESCO, Paris, 1978.

SVERDLOFF, S. N.: "The status of marine conservation in American Samoa", in: *Regional Symp. Conserv. Nat., Reefs a. Lagoons,* pp. 25–28, 1973.

TARDENT, P.: "Coelenterata, Cnidaria", in F. SEIDEL (ed.): *Morphogenese der Tiere 1.* Jena, pp. 69–415, 1978.

THOMASSIN, B. A.: "Feeding behaviour of the felt-, sponge-, and coral-feeder sea stars, mainly *Culcita schmideliana*", in: *Helgoländer wiss. Meeresunters. 28,* pp. 51–65, 1976.

VAUGHAN, T. W.: "The results of investigations of the ecology of the Floridian and Bahaman shoal-water corals", in: *Proc. Nat. Acad. Sci. Washington 2,* pp. 95–100, 1916.

VINE, P. J.: "Effects of algal grazing and aggressive behaviour of fishes *Pomacentrus lividus* and *Acanthurus sohal* on coral reef ecology", in: *Mar. Biol. 24,* pp. 131–136, 1974.

WELLS, J. W.: "Scleractinia", in: R. C. MOORE (ed.): *Treatise on Invertebrate Paleontology.* Pt. F. Geol. Soc. Amer., pp. 328–443, 1956.

WICKLER, W.: "Vergleichende Verhaltensforschung und Phylogenetik", in: G. HEBERER (ed.): *Die Evolution der Organismen I,* pp. 420–508. Stuttgart, 1967.

WIENS, H. J.: *Atoll environment and ecology.* New Haven, London, 1962.

WOOD-JONES, F.: *Coral and atolls.* London, 1910.

WRAY, J. L.: *Calcareous algae.* Amsterdam, Oxford, New York, 1977.

YONGE, C. M.: *A year on the Great Barrier Reef.* London, 1931.

YONGE, C. M.: "The biology of coral reefs", in: *Advances Mar. Biol. 1,* pp. 209–260, 1963.

YONGE, C. M.: "The nature of reef-building (hermatypic) corals", in: *Bull. Mar. Sci. 23,* pp. 1–15, 1973.

Sources of illustrations

(The figures refer to the consecutively numbered illustrations of the book.)

Bak, R. P. M., Curaçao: 41

Fricke, H. W., Seewiesen: 5, 24

Harmelin, J. G., Marseille: 2

Heckel, K.-G., Schwerin: 43, 50, 52, 60, 83, 95, 102, 107, 115, 123, 127, 133, 138, 139, 142, 143, 151

Jacana, Paris: 145, 147

Johnson, S., Eniwetok Atoll: 39, 59, 113, 116, 141

Kühlmann, D. H. H., Berlin: 9–12, 14, 16, 17, 19, 23, 25, 27, 28–31, 33–38, 40, 42, 45–47, 51, 55–58, 64–67, 69–71, 73–76, 80–82, 84, 85, 87, 89–93, 96, 108–112, 118, 121, 122, 124, 125, 130, 132, 135, 137, 148–150, 152

Playford, P. E., Perth: 72

Rabe, K., Schwerin: 1, 3, 4, 18, 22, 48, 49, 86, 97, 104, 105, 117, 119, 120, 126, 129, 131

Reese, E. S., Honolulu: 77–79

Schmied, H./Bavaria, Munich: 63

Schöne, H., Woltersdorf: 7, 8, 20, 21

Schuhmacher, H., Essen: 13, 26, 32, 144, 146

Strulik, D., Rostock: 53, 101, 103, 106, 114, 128, 136

Tschiesche, K.-H., Stralsund: 6, 15, 44, 54, 61, 62, 88, 94, 98–100, 134, 140

ZEFA, Düsseldorf: 68

If not otherwise mentioned in the captions to the line drawings, the models come from the author.

Index